PUBLICATIONS OF THE ISRAEL ACADEMY

OF SCIENCES AND HUMANITIES

SECTION OF SCIENCES

———

THE GENITALIA OF BOMBYLIIDAE (DIPTERA)

THE GENITALIA OF BOMBYLIIDAE

(DIPTERA)

by

OSKAR THEODOR

JERUSALEM 1983

THE ISRAEL ACADEMY OF SCIENCES AND HUMANITIES

ISBN 965-208-051-9

Printed at Keter Press, Jerusalem

CONTENTS

INTRODUCTION

DESCRIPTIONS OF Bombyliidae have been based in the past almost exclusively on external characters: wing venation and coloration, coloration of the integument and hair cover, form of the head, form of the antennae, absence or presence of pulvilli, etc. Many species have been described from single specimens and some of the above characters have proved to be very variable when a series of specimens was examined. This applies particularly to coloration which proved to be variable in specimens of the same species from different localities and often even in specimens of the same population. Descriptions of some species of *Exoprosopa* have been based on minor variations of the black pattern of the wings, which is also very variable.

The fine colour shades used by Austen (1937) and Efflatoun (1945) according to Ridgway, *Color Standards and Color Nomenclature* (1912), have proved of little value, as this work is usually not available at present to most entomologists and the name of the colours has little meaning without comparison with the colour plates. The extent of the black pattern on the integument varies considerably in the same species, as illustrated by Efflatoun (1945) for *Heterotropus* and *Mariobezzia*.

Wing venation also varies markedly, e.g., the position of the cross vein r-m, the closed or open cell r_5, and the presence of two or three submarginal cells.

The relative length and form of the segments of the antennae are much more variable than was assumed by the early authors, and the form of segment 3 is often changed by shrinking after the insect becomes dry. The form of the antennae was used by Bezzi (1924) and Engel (1935) to distinguish between the closely related genera *Anthrax* and *Spongostylum*. However, some species placed in *Anthrax* by Engel have an antenna which he described as being characteristic for *Spongostylum* (*A. aethiops* and related species).

The length of the proboscis is not a constant character, as it may be more or less extended to about twice its length from the retracted state.

Size is of little importance in many species. It may vary greatly in the same population. In some species there are two distinctly different forms — small specimens and much larger specimens — without transitional forms (e.g., *Bombylius ater* and *Thyridanthrax polyphemus*). This is probably due to development of the larvae in different hosts, and the small form has sometimes been described as a variety.

Some species have been considered to be highly variable and to have a very wide distribution, e.g., *Spongostylum ocyale*. This species was described from Nubia and has also been recorded from Egypt, Israel and Syria to Iran. Examination of the genitalia of specimens from Israel showed that they differ distinctly from the drawing of the genitalia given by Engel (origin of the specimen illustrated not stated), and there are at least two species with different genitalia identified as *S. ocyale* in the material from Israel.

The genitalia were little used by the early authors in the systematics of the family. The male genitalia are usually more or less retracted and covered by hairs, and Engel (1932, p. 4) thus stated that they are only very rarely useful for the distinction of species, as they are usually

1

covered by long hairs. Engel (1932-1937) gave drawings of the genitalia of a number of species — usually of the whole hypopygium in lateral view, and only in a few cases (*Villa, Hemipenthes*) detailed drawings of the aedeagus. Hesse (1938, 1956), in his monumental work on the Bombyliidae of Southern Africa, consistently described and illustrated the male genitalia of most species he examined. Drawings of the genitalia were also given by Zaitzev (1969), Painter (1963) and Hall (1975) but usually without sufficient details. Austen (1937) completely disregarded the genitalia and Efflatoun (1945) gave only a few drawings of their outer aspect in the complete insect. Hull (1973) gave a general description of the male genitalia and used modern terminology instead of the descriptive terminology of the earlier authors, e.g., Hesse. He gave illustrations of the genitalia of a representative of most genera, sometimes in several aspects, and also of some separate parts, but the labelling of the parts is incorrect in some cases (e.g., Figs. 844, 884). Hull usually illustrated the type species of the genus, which is often not typical of the genus, and disregarded the often great variation within a genus.

The female genitalia were completely disregarded by the earlier authors. Only recently, Mühlenberg (1970, 1971) described the differentiation of the posterior segments of the abdomen of the female and the spermathecae of a number of species. Marston (1970) illustrated the female genitalia of American species of *Anthrax*. It is amusing to note that Hesse gave illustrations of the single spermatheca of *Antonia* (1956, p.140) and of the spirally coiled spermathecae of *Systropus* (1938, p.1016) without recognizing them. He termed them coiled processes.

The use of the structure of the genitalia was found to be of particular importance in some genera and groups in which the external characters are very similar, e.g., *Geron* (see pp. 39–50) and the small black species of *Usia*. The female genitalia proved particularly valuable in these species, as they permit the identification of single females even if denuded or otherwise damaged. Distinctly different genitalia were found in specimens which apparently belonged to the same species according to external characters (*Petrorossia, fuscipennis* group of *Anthrax*).

The detailed structure of the genitalia of the Asilidae, particularly of the aedeagus and the spermathecae, has been described in a previous publication (Theodor, 1976). Examination of the genitalia made it possible to distinguish between externally very similar species and also showed that some species which are apparently well defined by external characters consist in fact of two or several different species (e.g., *Machimus setibarbus*). It also permitted determination of the systematic position in the subfamilies in a number of cases (*Trichardis, Cerdistus syriacus*).

Over 100 genera of Bombyliidae have been examined, including numerous species of many Palaearctic genera. Altogether, some 400 species were studied, about 200 of them collected in Israel, many in long series and including several new species which will be described in another publication. The material is deposited in the Department of Zoology of the Hebrew University, Jerusalem.

I am grateful to Prof. J. Kugler and his colleagues at Tel Aviv University for collecting much interesting material, including specimens from the Hermon area, the southern Negev and Sinai, by their technique of sweeping vegetation, thus obtaining long series of the very small species.

I am also grateful for material provided by: the Department of Entomology of the British Museum (Nat. Hist.); Dr D. J. Greathead, Commonwealth Agricultural Bureaux; Dr V. B. Whitehead, South African Museum, Cape Town; Dr. D. H. Colless, C.S.I.R.O., Canberra; and Dr. H. Schumann, Humboldt University, Berlin.

I am particularly indebted to Dr. J. C. Hall, University of California, who provided extensive material of American species which in some cases made it possible to determine whether genera recorded both from America and the Old World are in fact identical.

The drawings were made by Mrs. Ruth Chotiner and several by Dr. Joseph Schlein. Partial support was provided by the Fauna Palaestina Committee of the Israel Academy of Sciences and Humanities.

Mr. Robert Amoils and Miss Ilana Ferber edited the manuscript and prepared it for publication. The book was seen through the press by Mr. Shmuel Reem, Director of Publications Department, The Israel Academy of Sciences and Humanities.

TECHNIQUE

THE TECHNIQUE of preparation and staining of the genitalia has been described in detail in the publication on the genitalia of the Asilidae (Theodor, 1976). The aedeagus should be mounted in dorsal view and also in lateral view in some cases. The spermathecae should be stained with Haematoxylin if fresh material is available, as this permits the study of the gland and musculature.

MALE GENITALIA

The male genitalia of the Bombyliidae are usually small and much more uniform in structure than those of the Asilidae, but this applies only to the form of the gonopods and epandrium. The inner parts (aedeagus complex) are sometimes of simple structure, but have a complicated and specific form in many species. The genitalia are rotated through 180° in most genera examined, so that the gonopods are dorsal and the epandrium ventral (hypopygium inversum). They are partly rotated in some species and not rotated in some genera (*Eclimus, Systropus, Toxophora, Heterotropus, Conophorus* and others). The dististyli are usually folded back and not readily visible externally. Hesse (1938) considered the rotated position of the genitalia as normal and the non-rotated genitalia as exceptional (*Systropus*, p.994).

The epandrium (tergite 9) is undivided in all genera examined, with the exception of *Prorates* which, however, probably does not belong to the Bombyliidae at all, as discussed below (pp. 8, 18). It is of distinctive form in many genera and may have posterior processes and groups of setae. It is sometimes folded in half so that only its posterior margin, with a fairly narrow slit between the two halves, is visible, e.g., in *Empidideicus*. The cerci are usually short and broad and may bear dense, short, black spines in some species of *Anthrax*. They bear a characteristic raised area of dense black tubercles in the Ethiopian species of *Systropus*. The ventral part of the proctiger often contains sclerotizations of specific form, which were termed the subanal plate by Crampton (1942) and may be the vestigial sternite 10 or 11.

The gonocoxites (basistyli) are usually truncate-triangular and may bear characteristic processes and brushes of setae. The dististyli are usually situated apically and are of distinctive form in many species. Their exact form is best recognized in lateral view, as they are usually flattened laterally and therefore appear much narrower in dorsal view. In some species (*Cyrtisiopsis melleus*) they are situated in the middle of the gonocoxites. They are apparently absent or modified in some species of *Geron* and the apical part of the fused gonopods consists of processes of varying form. This part may also be folded back so that its exact structure is recognizable only in preparations. Some of these processes are probably homologous to the dististyli, since *Geron longiventris* and a species near *snowi* have two pairs of articulated processes, one of them closely resembling dististyli. There is often a triangular or rounded hypandrium or it is absent.

The aedeagus (Figs. 1–3) consists of a more or less long, conical tube, the sheath and an inner tube, the pump, a partly membranous pump chamber, and an apodeme. There are two basal plates of varying form, on which the protractor muscles of the apodeme originate. The aedeagus lacks distinct differentiations in the genus *Bombylius* and related genera but bears a process of specific form in many genera. This process is morphologically dorsal, situated between the aedeagus and epandrium, but it is situated ventral to the aedeagus in the natural position if the genitalia are rotated. This process is sometimes called the epiphallus but, since this term has been used for different structures in other

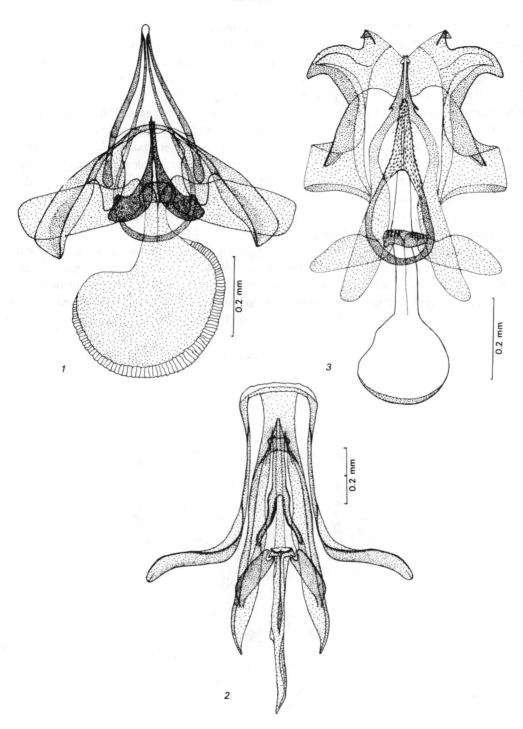

Figs. 1–3: Aedeagus
1. *Bombylius ater*; 2. *Exoprosopa aegina*; 3. *Petrorossia hesperus*

groups of insects, it is here called the aedeagal process. It is sometimes flattened and tongue-shaped or it may bear spine- or horn-shaped processes and is of specific form in practically every species examined. Its form is best recognized in dorso-ventral view. Drawings of the genitalia have usually been given in lateral view in which the differentiations of the aedeagal process are not recognizable.

The aedeagal process is of similar form in the species of some genera (*Villa, Exoprosopa*) but usually shows specific differentiations at the apex. The aedeagus is very long, narrow and curved in its basal part in *Amictus*. Only in two of the genera examined, *Cyllenia* and *Heterotropus*, is it divided into two or three long prongs and resembles the aedeagus of the *Machimus* group of the Asilidae. The aedeagus is connected to the lateral margin of the gonocoxites by two lateral struts (rami), and only its apodeme may project beyond their proximal margin. It is very large and its greater part is situated proximal to the gonocoxites in the genera *Platypygus* and *Empidideicus*.

A membranous tube with 1–3 sclerotized ridges attached to the head of the aedeagal apodeme has been found so far only in the Anthracinae (Figs. 4a, b). This is apparently an 'endoaedeagus' which has been described for the Asilidae in which it is present in many genera. It is covered with denticles and platelets in the Asilidae, but not in the Anthracinae. There is a similar structure in *Lomatia* and related genera.

A distinctly different aedeagus is present in the genus *Prorates* (*Alloxytropus*) (Figs. 5, 6). The aedeagus is divided into two prongs which end in long, thin, coiled tubes. The

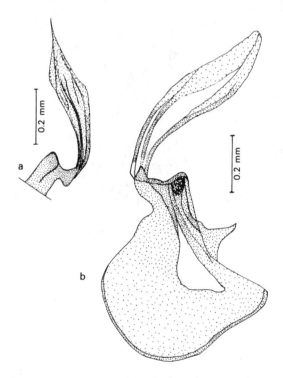

Fig. 4: Endoaedeagus
(a) *Anthrax* sp.; (b) *Chionamoeba* (?) *lepida*

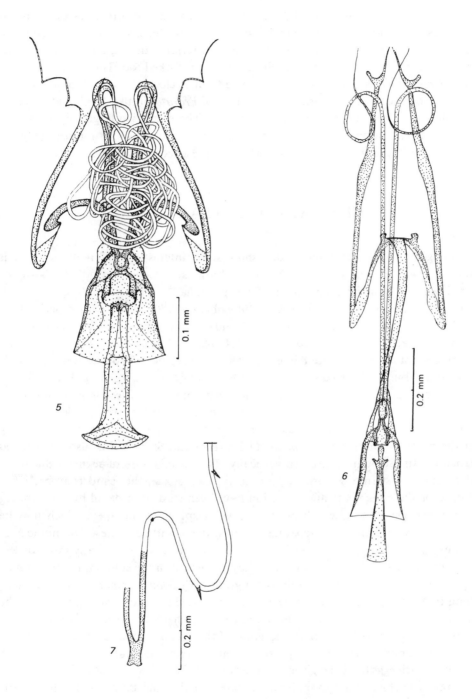

Figs. 5–7: Aedeagus
5, *Prorates anomalus*; 6. *Prorates frommeri*;
7. tube of aedeagus of *Prorates* sp. with spines

aedeagus is articulated at its base with a fork-shaped rod which originates on a curved frame. At rest, this rod is directed anteriorly, so that the aedeagus is deeply retracted. During copulation this rod is apparently turned posteriorly through 180°, so that the aedeagus is extended posteriorly by twice the length of the forked rod. This structure differs markedly from that in all other genera examined but it closely resembles that in some genera of Scenopinidae (e.g., *Belosta*); the position of *Prorates* in the Bombyliidae is thus questionable (see below under Classification, p. 18). There are two large spines on each tube of the aedeagus in one species examined, the function of which is not clear (Fig. 7). This two-pronged aedeagus is associated with the presence of two spermathecae.

FEMALE GENITALIA

The female genitalia of the Bombyliidae show some interesting differentiations. It is surprising to find sometimes a mass of fine sand grains in a concavity of the posterior abdominal segments during preparation of the genitalia.

This phenomenon has been studied by Mühlenberg (1970, 1971) who described an 'oviposition apparatus' and the formation of a 'sand chamber'. This apparatus is present in all Tomophthalmae other than *Mariobezzia* and *Antonia*, but is absent in some subfamilies of the Homoeophthalmae. Tergite 8 is invaginated into segment 7 so that it is not visible externally and forms a nearly complete ring with a narrow ventral opening. It usually has an anterior apodeme on which two strong muscles are inserted. These muscles move the tergite up and down and also laterally by the action of one of the two muscles. Tergite 8 usually bears a more or less dense brush of hairs at the posterior margin. It bears strong setae at the posterior margin in *Eclimus* and *Thevenemyia*. Sternite 8 is also invaginated into segment 7 and sometimes situated vertically. The invagination of segment 8 and the vertical position of sternite 8 form a posteriorly open space, the 'sand chamber'. The musculature of these segments and its action have been studied in detail by Mühlenberg (1971). In many genera tergite 9 bears two rows of spines at the apex, which may be situated on partly or completely separate sclerites, the acanthophorites. The number of these spines varies widely, from 3–4 in a row to over 30. They are usually more or less spatulate, with a curved, hook-shaped end. They are absent in the subfamilies placed here in group 1 and also in some genera of the other groups (*Cyllenia, Antonia, Lordotus* group). According to Mühlenberg, the spines on tergite 9 are used to scrape sand grains from the soil and the brush on tergite 8 moves them into the sand chamber. During oviposition, the eggs, which are coated with the sticky secretion of the accessory glands, become covered with the sand grains, perhaps to protect them against damage or desiccation. A photograph of such eggs has been given by Hull (1973, p. 37).

The spermathecae (Fig. 8) are more highly differentiated in the Bombyliidae than in the Asilidae. They consist of a sperm reservoir or capsule and a duct leading to an ejection apparatus provided with muscles from which extend ducts to the dorsal wall of the vagina, where they open inside a sclerotized frame, the furca. The capsule is partly or completely surrounded by a gland consisting of cells with intracellular canaliculi which open into the

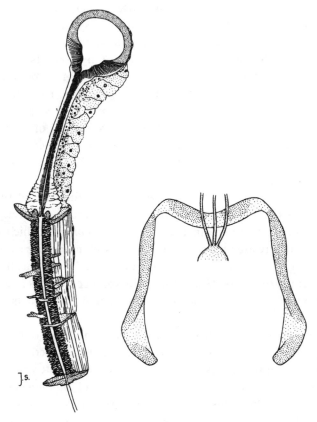

Fig. 8: *Bombylius* sp. (near *venosus*), spermatheca

lumen of the capsule. This gland sometimes extends onto the duct. The form of the sperm capsule varies considerably. It may be globular, cylindrical, pear-shaped or sausage-shaped and is usually more or less strongly sclerotized. It may also be tubular and the tubes sometimes form a spiral (*Apolysis, Toxophora, Systropus, Doliopteryx*) or a dense coil (*Geron*). All these forms are also present in the Asilidae but a dense coil, which has been found to date only in *Geron* of the Bombyliidae, is common in the Asilidae and so is a spiral. The ducts are sometimes very long, extending through the greater part of the abdomen, but the spermathecae are usually situated in the 2–3 posterior segments, particularly if the eggs are developed. The ducts sometimes show characteristic widenings in the middle or in their proximal part. The ejection apparatus forms a sclerotized part of the duct with a rounded plate at one or both ends and short and longer processes along the duct. Muscle fibres extend between and from the plates to the processes on the duct. The length of the ejection apparatus varies greatly; it is usually short, but very long in some genera (*Amictus, Sinaia, Cytherea*). It is apparently absent in some genera (*Spongostylum, Prorachtes*). There is a special development of the ejection apparatus in some genera of Exoprosopinae. The basal end plate is transformed into a transparent, funnel-shaped structure (Figs. 729, 768). Among the Old World species this has been found only in

Hemipenthes but it is present in a number of American genera of Exoprosopinae. The ducts from the ejection apparatus to the vagina are sometimes sclerotized at the end and they may form a common duct before opening into the vagina. The common duct may be sclerotized or wide and membranous. The furca usually consists of two separate bars connected by a membrane, and there is sometimes a sclerite of characteristic form between their anterior ends. It is U- or V-shaped in some genera of Exoprosopinae and shows specific differences in many species. There are complicated sclerites of specific form behind the furca in some genera. These sclerites have proved of particular importance in *Geron* in which the sperm capsules are very similar in all species of the genus. They permit the identification of single females which are difficult to determine by external characters.

There are sclerotizations and membranous widenings at the base of the sperm capsule, which are especially well developed in species of *Exoprosopa* and related genera.

All genera examined have three spermathecae, except *Antonia* in which there is only one, and *Prorates* in which there are only two. The median spermatheca is rudimentary in a new genus of Mythicomiinae. The three spermathecae are usually similar, but the median one may be much longer (*Cyrtisiopsis, Glabellula*) or its sperm capsule is much larger than that of the lateral spermathecae (*Spongostylum mixtum*) and their ducts are partly deeply constricted.

There is a special development in a few genera. The proximal duct of the spermatheca does not open directly into the vagina but forms an open, funnel-shaped tube which opens in a basal, membranous widening or reservoir; from this a duct extends to the vagina. This has been found so far only in the genera *Glabellula* and *Doliopteryx* of the subfamily Mythicomiinae (Figs. 43, 46). This reservoir may serve to store the surplus sperm which cannot be contained in the usually small sperm capsule. A similar differentiation with a funnel and a reservoir, but of more complicated structure, is present also in the genus *Belosta* of the family Scenopinidae. An ejection apparatus with muscles is absent. This reservoir was labelled the spermatheca by Kelsey (1969), who failed to find the globular sperm capsule.

A similar reservoir but combined with a more highly developed ejection apparatus with muscles has been described by Berlese (1909) for Pentatomidae (*Graphosoma*) and by Pendergrast (1957) also for other genera of the family. It is surprising that such a highly specialized apparatus should be present in two such widely remote groups of insects. Weber (1933) gave diagrams of the female reproductive organs of 12 orders of insects but indicated such a differentiation combined with a muscular ejection apparatus only for Pentatomid bugs. He wrongly labelled this reservoir as the vagina.

The spermathecae are sometimes similar in all species of a genus (*Usia, Geron, Phthiria*), but there are spermathecae of different forms in some genera which appear to be well defined by external characters and the male genitalia (*Anastoechus*—2 types, *Amictus*—3 types, *Anthrax*—2 types in the Old World species).

The situation is particularly complicated in the genus *Exoprosopa*. This genus has been divided into numerous subgenera on the basis of the venation and pattern of the wings. Six different types of spermathecae have been found in the 30 species of the genus examined: 10 had rounded capsules with membranous differentiations at their base (type 1), 10 had

10

short, tubular, conical capsules (type 2), and 4 other types of spermathecae were found, each in one or two species. The grouping of the species according to the type of the spermathecae does not agree at all with that based on wing venation and pattern. The spermathecae show distinct specific differences in some groups of *Bombylius* which are well defined by external characters (*ater* group).

The structure of the posterior abdominal segments of the female varies distinctly in the various subfamilies. In *Bombylius* and related genera tergite 9 is moderately wide, curved, with long proximal processes. The acanthophorites are long and curved. The arrangement is similar in *Heterotropus*, but the acanthophorites are short and wide. In the Anthracinae and related subfamilies the arrangement is similar, but tergite 9 is very narrow and curved. The acanthophorites are long, narrow and curved.

The structure of these sclerites is different in some subfamilies. In the Corsomyzinae, tergite 9 is transversely rectangular and the spines are situated at the posterior margin of the tergite. In the Exoprosopinae, tergite 9 is large, truncate-conical and the spines are situated at the posterior lateral corners of the tergite, sometimes on a tubercle (Figs. 9 – 12).

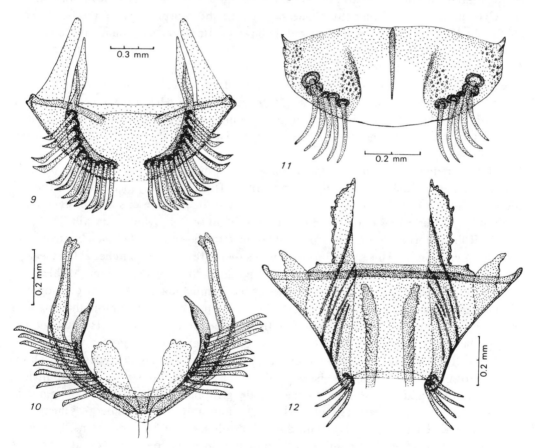

Figs. 9–12: Tergite 9 of female
9. *Bombylius cruciatus*; 10. *Anthrax fuscipennis*;
11. *Mariobezzia lichtwardi*; 12. *Exoprosopa deserticola*

CLASSIFICATION

THE FAMILY Bombyliidae contains very different forms—very small flies (1-3 mm) with reduced wing venation to very large flies with a complicated wing venation and a very specialized morphology of the head. There is practically no definite diagnosis of the family. The classification within the family is controversial. Bezzi (1924) divided the family into two groups: the Homoeophthalmae in which the posterior margin of the eyes is complete and R_{2+3} originates at an acute angle from R_{4+5}, far proximal to the cross vein r–m; and the Tomophthalmae in which there is an indentation of the posterior eye margin from which extends a horizontal bisection line, the occiput is deeply invaginated, and R_{2+3} usually originates vertically at or near the cross vein r–m. This division has been accepted by most later authors. The two groups have been differently defined by Hesse who placed the Cylleniinae in the Tomophthalmae because of the morphology of the head. This subfamily was placed in the Homoeophthalmae by Bezzi on account of the complete posterior margin of the eyes. Its position in the Tomophthalmae was accepted by Hull (1973) and has also been adopted here.

The family has been divided into numerous subfamilies (20–23) by various authors. Some of these subfamilies contain only a single genus, e.g., the Heterotropinae. The other genera which Melander (1950) placed in this subfamily, and which was accepted by Hull, are related to *Prorates* and do not belong to the Heterotropinae or, apparently, to the Bombyliidae at all.

Hull (1973) reduced the number of subfamilies to 12, made some of them tribes of other subfamilies, and established a number of new tribes. This is certainly justified but certain of his tribes appear to be insufficiently substantiated and the position of some of them in the subfamilies seems to be doubtful, e.g., in the subfamily Bombyliinae, as will be shown below. Rohdendorf (1964) raised some subfamilies to the rank of families (Systropinae, Usiinae, Cyrtosiinae). This seems inconsistent as there are other subfamilies at least as well characterized and distinct, which were not upgraded to families by him. Mühlenberg (1971), in his phylogenetic systematic studies of the Bombyliidae, rejected the division into Homoeophthalmae and Tomophthalmae. He maintained only the Tomophthalmae as a monophyletic group but left the position in the Homoeophthalmae open. He introduced a new element in his phylogenetic scheme—the differentiation of the posterior abdominal segments of the female into an oviposition apparatus which has been described above (p. 8). However, his scheme is based only on a small number of species, mainly Mediterranean, and is restricted to the female genitalia, although the male genitalia probably also contain characters of phylogenetic importance. Mühlenberg's scheme has been discussed by Bowden (1974) who disagrees with some of his conclusions. However, the introduction of a new complex of characters into phylogenetic considerations is certainly a valuable contribution.

If all the above-enumerated characters (form of the head, indentation of the posterior eye margin, wing venation, differentiation of the posterior abdominal segments of the female)

12

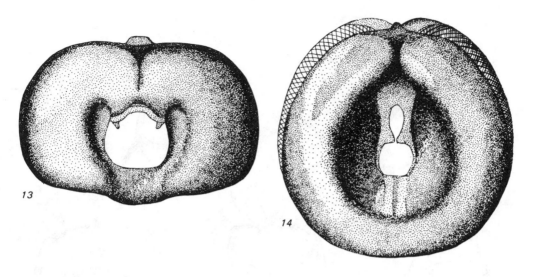

Figs. 13–14: Head, posterior view
13. *Bombylius* sp.; 14. *Thyridanthrax telamon*

are considered, it appears that there is only a single complex of characters which distinguishes the Tomophthalmae from the Homoeophthalmae, i.e., the form of the head, with a deeply invaginated occiput and two occipital foramina, described in detail below (Figs. 13–19). This invaginated occiput is present in all subfamilies of the group and supports Mühlenberg's view that the Tomophthalmae are a monophyletic group. None of the other characters are associated exclusively with the form of the head of the Tomophthalmae; they either occur sporadically in other groups (e.g., the indentation of the posterior eye margin in *Efflatounia* and related genera) or they form a series of transitions. Thus, the origin of R_{2+3} is similar in the Homoeophthalmae and in some subfamilies of the Tomophthalmae, as is the differentiation of the posterior abdominal segments of the female.

There are apparently three main groups in the family, with transitional forms between them.

Group 1 consists of subfamilies in which the occiput is normal, either flat, slightly convex (Fig. 13) or more or less bulging or prolonged. There is a single occipital foramen, no indentation of the posterior eye margin, the wing venation is either reduced (one or two radial veins and sometimes absence of a discal cell) or normal (four branches of the radius and three branches of the media), and the posterior segments of the female are not differentiated. Tergite 8 is not invaginated and neither is sternite 8. There are no spines on tergite 9. There is thus no sand chamber. This group contains the subfamilies Mythicomiinae (Cyrtosiinae), Toxophorinae, Usiinae, Systropinae and Phthiriinae.

The subfamily Geroninae agrees in most characters with the above subfamilies of the

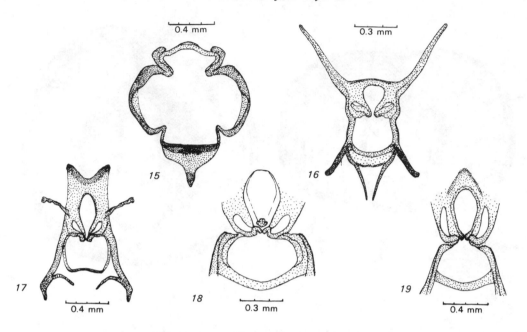

Figs. 15–19: Occipital foramina
15. *Bombylius discoideus*; 16. *Mariobezzia lichtwardi*;
17. *Cytherea syriaca*; 18. *Spongostylum ocyale*; 19. *Exoprosopa grandis*

group, but shows incipient differentiation of tergite 8 and incipient invagination of sternite 8 of the female. However, there are complex structures behind the furca of the female, and the male genitalia are highly differentiated. The gonocoxites are fused and apically bear a complex group of processes of varying form. Most species lack articulated dististyli, but in two species there are two pairs of partly articulated processes, one of which is probably homologous to the dististyli. This complex of processes can be folded back. Tergite 9 bears no spines and a sand chamber is absent. The male genitalia differ distinctly from those of most other Bombyliidae examined, but show some similarities to those of the Toxophorinae.

The subfamily Heterotropinae is here considered to contain only the genus *Heterotropus* and a new genus or subgenus which will have to be established for the Nearctic species *H. senex*. The species exhibit most of the characters of the above subfamilies but they have partly separated acanthophorites with spines and the cerci are long, usually folded back and bear hairs and setae. Sternite 8 is not invaginated and there is thus no sand chamber; the presence of spines on tergite 9 is thus not necessarily connected with the collection of sand. The biology of the species is not known but most specimens were collected resting on loose sand and they probably oviposit in the soil which is loosened by the spines. Tergite 8 of the female is partly invaginated; it has no anterior apodeme but its anterior margin is strongly sclerotized and there is a median longitudinal ridge in its posterior half. The male genitalia differ distinctly from those of the other genera examined, except the Cylleniinae, in the long, three-pronged aedeagus and in the large dististyli of characteristic form.

14

Systropus also shows incipient differentiation of the posterior segments of the female. Tergite 8 is invaginated and has an anterior apodeme which is directed dorsally. Sternite 8 is also invaginated but does not form a sand chamber. There are no spines on tergite 9. The male genitalia are not rotated. The epandrium has long posterior lateral processes, the gonocoxites are fused, and the dististyli are of specific form. The aedeagal process varies in form and there are additional processes and sclerites on the aedeagus, the relationship of which is not clear.

The subfamily Ecliminae was placed as a tribe in the subfamily Bombyliinae by Hull. This seems to be incorrect, as the Ecliminae differ in most important characters from the Bombyliinae: form of the head and wings, differentiation of the posterior segments of the female, and others.

The male genitalia are not rotated. Tergite 8 of the female is only partly invaginated, strongly sclerotized, with hairs in its posterior half, clearly visible externally and with strong spines and setae at the posterior margin. It has an anterior apodeme and is thus apparently movable. Sternite 8 of the female is also not invaginated, externally visible, strongly sclerotized, convex posteriorly and bears long hairs. Segment 8 is usually partly retracted into segment 7. Tergite 9 small, rhomboidal. Acanthophorites with two rows of strong, spatulate spines with a hook-shaped end which become larger laterally.

Mühlenberg (1971, p.31) assumed that there is 'a kind of sand chamber which differs from that of the other species described'. This seems to be not correct. Examination of specimens showed that there is apparently no sand chamber and the differentiation resembles that in *Systropus*. The armature of setae and spines on tergites 8 and 9 is apparently a specific development, probably connected with oviposition in the soil.

The last four subfamilies thus show characters transitional between groups 1 and 2.

Group 2 consists mainly of the subfamily Bombyliinae. Hull placed the subfamilies Cythereinae and Mariobezziinae as tribes in this subfamily, but they belong to the Tomophthalmae, as will be shown below. They have an invaginated occiput and two distinctly separated occipital foramina (Figs. 16, 17).

The occiput of the species of this group is not invaginated and there is no indentation or bisection line at the posterior eye margin, except in *Efflatounia* and related species.

The occipital foramen of all Homoeophthalmae examined is a single opening of varying form, but it shows incipient division by two short processes into a smaller dorsal and a larger ventral part. These processes are articulated with rod-shaped cervical sclerites. There are additional processes projecting into the ventral part of the foramen in a few species (*Acanthogeron, Efflatounia*). The foramen of *Cyrtisiopsis melleus*, which is situated at the end of the prolonged occiput, is of specific form. A similar, more complicated, incomplete division of the foramen was found also in some Cyclorrhapha.

The wing venation resembles that in group 1, but the radius always has four branches. There is a sand chamber in the female and tergite 8 is invaginated and has an anterior apodeme and a usually well-developed brush of hairs at the posterior margin. Sternite 8 is invaginated and often situated vertically and there are acanthophorites with spines. *Lordotus* and *Geminaria* show a completely different development of the abdomen (see below, pp. 130–133).

15

The male genitalia either lack special differentiations or there are complicated differentiations of the aedeagal process (*Bombylisoma, Efflatounia*). *Efflatounia* clearly belongs to the subfamily Bombyliinae but it has an indentation of the posterior eye margin. A deep, rounded invagination of the posterior eye margin is present also in *Eurycarenus* and other genera of the tribe Heterostylini established by Hull.

Group 3 consists of the subfamilies Corsomyzinae, Cylleniinae, Cythereinae, Lomatiinae, Anthracinae and Exoprosopinae. The main character of this group is the form of the occiput (Fig. 14). It consists of an outer, ring-shaped, convex part which forms dorsal lobes, leaving only a narrow cleft behind the ocellar tubercle. A median dorsal sclerite (cerebrale) is absent. The greater inner part of the occiput is invaginated like a funnel so that the occipital foramina are situated far anteriorly in the middle of the head. The head is thus attached only in a very small area, which accounts for its great mobility (but also causes its frequent loss in collections). The occipital foramen is divided into a larger, rounded or nearly square ventral opening through which the intestinal tract passes and a smaller, dorsoventrally oblong opening through which the aorta and tracheae probably pass. The dorsal opening is present in all genera of Tomophthalmae examined but differs in size and form from genus to genus (Figs. 15–19). It has been illustrated by Hull (1973, Figs. 748, 752) but is not mentioned in the text. There are two additional, small, oblong openings lateral to the base of the dorsal opening in most species. This is a unique, highly specialized formation which obviously developed in connection with the invagination of the occiput. The presence of this structure in the Cylleniinae, Corsomyzinae and Cythereinae clearly proves their close relationship to the Tomophthalmae.

The wing venation varies. In the Cylleniinae, Corsomyzinae and Lomatiinae, the origin of R_{2+3} resembles that in the Homoeophthalmae, forming an acute angle far proximal to the cross vein r–m. It originates closer to the cross vein r–m, more vertically and curved in the Cythereinae and in *Petrorossia*. In the Anthracinae and Exoprosopinae, it originates at or very near the cross vein r–m, vertically, often forming a right angle and with a small process.

An indentation of the posterior eye margin is present in the Lomatiinae, Anthracinae and Exoprosopinae and in *Antonia* but is absent in the other three subfamilies. A horizontal bisection line is present in some genera, absent in others.

The differentiation of the posterior segments of the female resembles that in group 2 of the Homoeophthalmae in all subfamilies, except the Exoprosopinae. The sand chamber in this subfamily is formed mainly by sternite 8 which has become membranous, except for its posterior margin, and forms a large, hemispherical, posteriorly open cavity. Tergite 8 is invaginated; it either has a long, anterior apodeme, a short, wide apodeme or an apodeme is absent and the muscles are inserted at its anterior margin. The posterior brush of hairs is often reduced and sparse.

The subfamily Cylleniinae was placed in the Homoeophthalmae by Bezzi and other authors because there is no indentation of the posterior eye margin. However, Hesse and Hull placed it in the Tomophthalmae because of the morphology of the head, and this seems correct, since the occiput is invaginated and there are two occipital foramina. The wing venation resembles that of group 2 of the Homoeophthalmae, with R_{2+3} originating in

16

an acute angle far proximal to the cross vein r – m. The sand chamber resembles that of the Bombyliinae. There are separate acanthophorites with spines in *Amictus* but spines are absent in *Cyllenia*. The aedeagus is long and narrow, divided into three prongs in *Cyllenia*, but undivided in *Amictus*. The subfamily Tomomyzinae, of which 3 American genera were examined, conforms in all important characters with the Cylleniinae and is here considered as a tribe of this subfamily following Hull (1973).

The genus *Mariobezzia* of the subfamily Corsomyzinae was placed by Bezzi (1924) in the subfamily Mariobezziinae. Hull placed it as a tribe in the subfamily Bombyliinae and stated that it should probably be united with his tribe Corsomyzini. Bowden (1975a) showed that *Mariobezzia* closely resembles some genera of Corsomyzinae (e.g., *Gnumyia*) in the form of the head and in the male genitalia. This is confirmed by the illustrations of the male genitalia of the *Corsomyza* group given by Hesse (1938) and by examination of 3 genera of the group. The two tribes should thus be united and are placed here in the subfamily Corsomyzinae of the Tomophthalmae, as *Corsomyza* was the first genus described by Wiedemann in 1820.

The head of the species of the group shows a very pronounced development of the face. The occiput is invaginated and there are two distinctly separated occipital foramina. The origin of R_{2+3} resembles that in the Cylleniinae. Aedeagus simple, without an aedeagal process and with a narrow, curved apical part. Tergite 9 of the female with strong spines, but acanthophorites are not separated. Sternite 8 normal, membranous in its greater part and not invaginated. There is thus no sand chamber. Tergite 8 is of characteristic form, with a more or less broad apodeme (Figs. 423, 425) and a sparse brush of hairs posteriorly. Spermathecae simple, with long ducts and small, more or less narrow, oblong-oval or cylindrical capsules.

The subfamily Cythereinae was placed by all authors in the Homoeophthalmae, but Engel (1935, p. 319) already stated that it is transitional to the Tomophthalmae in its wide frons in both sexes and the widely separated antennae. Hull (1973, pp. 69, 150) also noted a resemblance between this subfamily and the Tomophthalmae in some characters but considered this as due to convergence. The wide frons is not a character only of this subfamily and the Tomophthalmae; a similar wide frons is present also in some genera of the Bombyliinae (*Anastoechus, Acanthogeron, Bombylodes*) and in both sexes in some species of *Usia*.

The occiput is deeply invaginated, with two distinctly separated occipital foramina. The dorsal foramen is connected with the ventral foramen by a narrow slit. The origin of R_{2+3} resembles that in *Petrorossia*, being curved, not at an acute angle, more or less far proximal to the cross vein r – m (1.5 to 5 times the length of r – m). The aedeagal process has a broadened end and shows specific differences. Sand chamber as in the Bombyliinae, Lomatiinae and Anthracinae. The spermathecae are simple, with very long ducts. Ejection apparatus also very long, capsules oblong-oval. Acanthophorites long, curved, with broad proximal part.

The Nearctic genus *Pantarbes* resembles *Cytherea* but differs from it in some characters. The antennae are situated much more closely together. The dorsal occipital foramen is oblong-oval and more broadly connected with the ventral foramen. The spermathecae resemble those of *Cytherea* in general, but are much shorter. The ejection apparatus is also

17

much shorter and more strongly sclerotized. The dististyli are quite different from those of the species of *Cytherea* examined and resemble those of *Dischistus*. All these characters distinctly distinguish the Cythereinae from the other subfamilies of the Tomophthalmae. The subfamily is therefore maintained here.

The Lomatiinae are transitional in some characters between groups 2 and 3. The head is typical for the Tomophthalmae, deeply invaginated and with two occipital foramina. There is an indentation of the posterior eye margin, but a bisection line is either very short or absent. The wing venation resembles that of the Homoeophthalmae in the origin of R_{2+3}. The sand chamber is similar to that of the Bombyliinae, but the acanthophorites are not separated and the spines are situated on tergite 9. The aedeagus may be simple, without an aedeagal process (*Lomatia*), or may have an aedeagal process with complicated differentiations (*Petrorossia*) or more or less complicated apical processes.

The Anthracinae have the typical head and wing venation of the Tomophthalmae but the sand chamber resembles that of the Bombyliinae. The aedeagal process shows specific differentiations in every species examined. The presence of an 'endoaedeagus' (see above, p.6) in all genera of the subfamily examined is a special character of the subfamily. The spermathecae vary distinctly in the different genera and provide valuable specific and generic characters. Acanthophorites separated, long, narrow, curved.

The Exoprosopinae are distinguished by the special development of the posterior segments of the female and by the considerable variation of the spermathecae. Tergite 9 large, truncate-conical, with spines. Acanthophorites not separated.

The genus *Antonia* is difficult to place in any of the subfamilies described. Hull placed it as the tribe Antoniini in the Lomatiinae. It has the typical head of the Tomophthalmae, with a deeply invaginated occiput, two occipital foramina, an indentation of the posterior eye margin, and a bisection line. The origin of R_{2+3} resembles that in the Homoeophthalmae. The abdomen of the female is completely undifferentiated, long and narrow, with 9 externally visible segments. Tergite 9 lacks spines and its only differentiation are two long appendages in some species. This genus is also unique in having only a single spermatheca. The male genitalia resemble those of other Tomophthalmae, with a flattened, tongue-shaped aedeagal process. The apodeme of the aedeagus has two broad lateral flanges so that it is cruciform in transverse section. This is such an aberrant combination of characters that this genus may deserve subfamily status as suggested by Hesse (1956, p.50), at least as much as other monogeneric subfamilies.

The genus *Prorates* (*Alloxytropus*) cannot be placed in the above scheme. It was described originally as an Empid and later placed in the subfamily Heterotropinae by Melander (1927, 1950), but this is certainly incorrect. The resemblance to *Heterotropus* is superficial. The head and the male genitalia are completely different (see above, p. 12). The aedeagus closely resembles that of a species of Scenopinidae (*Belosta viticolapennis*) as illustrated by Kelsey (1969, p.282). The spermathecae also differ distinctly from those of all other Bombyliidae examined. There are only two spermathecae. The sperm capsules are broadly cup-shaped, membranous, with a sclerotized apical rim. The ducts show complicated differentiations in some of the species examined (Figs. 20–22).

Prorates also resembles the Scenopinidae in the presence of a sensory area with short spines

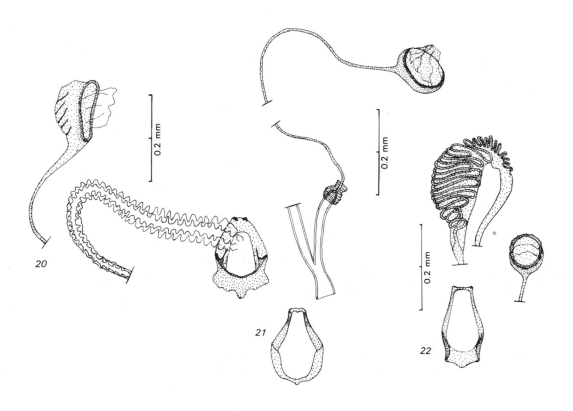

Figs. 20–22: Spermatheca
20. *Prorates* sp. no. 1; 21. *Prorates frommeri*; 22. *Prorates* sp. no. 2

and hairs on tergite 2 of the abdomen (Fig. 23a, b). Such sensory areas, usually two, are present in all Scenopinidae, as Dr. Kelsey informed me (in litt.), but they are absent in all Bombyliidae examined. Several authors have remarked on the resemblance of *Prorates* to the Scenopinidae (Engel, 1933; Efflatoun, 1945) from which it differs mainly in the forked vein M_{1+2}. This genus probably does not belong to the Bombyliidae at all and it would be more correct to place it in the Scenopinidae, as the subfamily Proratinae. This has also been suggested by Hull (1973, p. 228) who stated: 'It is just possible that the relationship [to the Scenopinidae] is more than superficial'.

The series of transitions described should not be considered as a phylogenetic scheme, and the genera in group 1 should not be considered as primitive since some of them show distinct, complicated specializations of the genitalia, e.g., in the male genitalia of *Geron* and *Toxophora* and in the spermathecae of *Glabellula* and *Doliopteryx*.
Bowden (1974) stated that the Homoeophthalmae apparently contain a number of groups which developed independently, and an apparently primitive condition may not be the plesiomorph homologue of the apomorph condition in the Tomophthalmae.

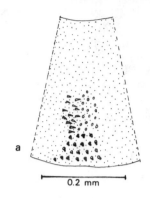

0.2 mm

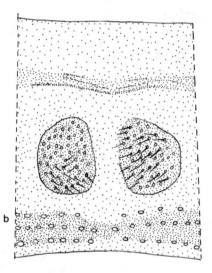

Fig. 23:   Sensory areas on abdominal tergite 2
(a) *Prorates frommeri*; (b) *Scenopinus fenestralis*

SYSTEMATIC PART

Divison *HOMOEOPHTHALMAE Bezzi*

MYTHICOMIINAE (CYRTOSIINAE) Melander, 1902

THIS SUBFAMILY contains very small forms with reduced wing venation. However, the species are not primitive, as the genitalia show distinct, complicated differentiations in certain species, some of which have not been found in other Bombyliidae, e.g., the spermathecae of the genera *Glabellula* and *Doliopteryx*.

Empidideicus Becker, 1907

E. carthaginiensis, mariouti, and several undescribed species (Figs. 24–31)

Epandrium broad, short, rounded, with posterior lateral corners, folded in half in life, so that only a narrow space remains between the two halves. Gonocoxites short, broadly

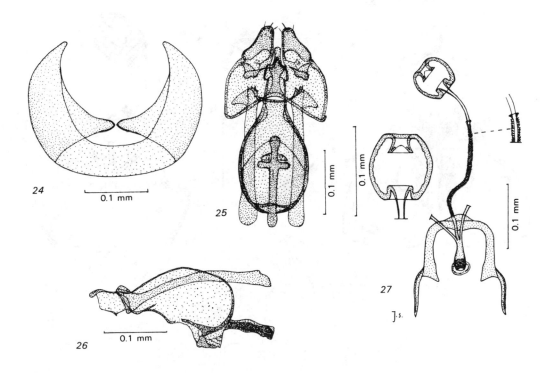

Figs. 24–27: *Empidideicus carthaginiensis*
24. epandrium; 25. gonopods and aedeagus;
26. aedeagus, lateral; 27. spermatheca

21

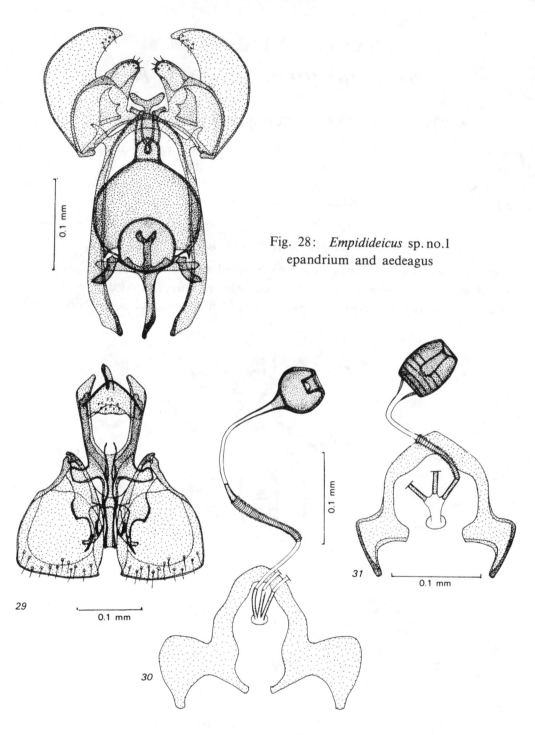

Fig. 28: *Empidideicus* sp. no. 1
epandrium and aedeagus

Figs. 29–31: *Empidideicus mariouti*
29. gonopods and aedeagus; 30. spermatheca type 1; 31. same, type 2

triangular. Dististyli short, triangular, with rounded apex. Aedeagus very large, bulbous in the basal part which extends distinctly proximal to the gonocoxites; its apex wide, conical, sclerotized. Aedeagal process not projecting apically, consisting of two long rods connected by a transverse bar. Apodeme with two lateral processes near its head. The aedeagus of *mariouti* and of an undescribed species show additional differentiations.

Spermathecae with globular, slightly flattened or barrel-shaped capsules with an apical indentation of varying form, short ducts and weakly sclerotized, striated ejection apparatus. Capsules with specific differences. Furca U- or V-shaped, with specific differences.

The species are difficult to distinguish on the basis of external characters. Two females resembling *mariouti* in external characters had distinctly different spermathecae (Figs. 30, 31).

Cyrtosia Perris, 1839

C. cognata, (?) *opaca* and several unidentified species (Figs. 32–38)

Epandrium (*cognata*) with two posterior processes and two inward directed, pointed

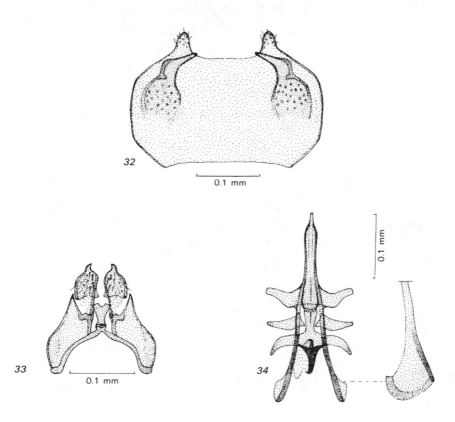

Figs. 32–34: *Cyrtosia cognata*
32. epandrium; 33. gonopods; 34. aedeagus

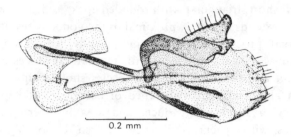

Fig. 35: *Cyrtosia* sp. no. 1, genitalia, lateral

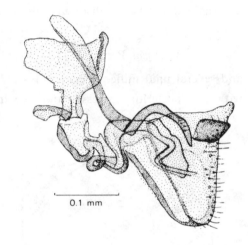

Fig. 36: *Cyrtosia* sp. no. 2, genitalia, lateral

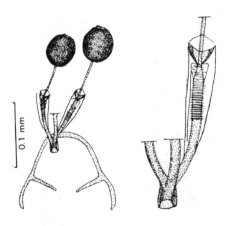

Fig. 37: *Cyrtosia* (?) *opaca*, spermathecae

Fig. 38: *Cyrtosia* sp. no. 3, spermatheca

processes. Gonopods small; gonocoxites triangular; dististyli wide, with pointed apex. Aedeagus narrow, conical. Aedeagal process not projecting beyond apex of aedeagus, with lateral processes and long posterior processes with widened end. In another species (species no. 1) the basal part is situated at a right angle to the apical part. In a yet another species (no. 2) the apical part is twisted into an irregular spiral.

Spermathecae with globular, ovoid or pear-shaped capsules with a reticulate pattern in the apical part. In species no. 3 the capsules are completely reticulated, with a cup-shaped structure at the beginning of the ducts. The ducts are usually little differentiated, but in (?) *opaca* there is a distinct, complicated ejection apparatus. Furca U- or V-shaped, with large inner processes and specific differences. Basal part and common duct of *C. cognata* sclerotized.

Glabellula Bezzi, 1902

G. nobilis and two undescribed species (Figs. 39 – 44)

The genitalia of both sexes are highly differentiated and show, particularly in the female, differentiations not found in other Bombyliidae, except in the related genus *Doliopteryx*.

Epandrium rounded, with two long, inward directed, black processes which are visible externally. Gonocoxites short, broad, fused, with two pointed processes in the middle of the posterior margin. Dististyli of irregular form, wider apically. The aedeagus of species no. 1 is conical, with a slightly widened, sclerotized apex. Aedeagal process not reaching apex of aedeagus, with a broad, rectangular apical part with rounded margin and two openings. Apodeme large, with wide, flattened, lateral processes near the apex. In species no. 1 the sheath has two pointed, slightly curved, black, apical processes.

The spermathecae have small, conical capsules. Ducts long, sclerotized in their apical part. They have a tripartite sclerotization in the middle, the wider median part of which is reticulated. This is possibly a displaced, additional sperm capsule. The apical capsules are very small. The ducts continue into an open funnel which opens into a wide, membranous reservoir from which extends a duct to the vagina. The median spermatheca has a very long funnel and a large reservoir. The lateral spermathecae are similar, but the funnel is much shorter and the reservoir small. Furca V-shaped, with two large openings at the apex through which the lateral spermathecae open.

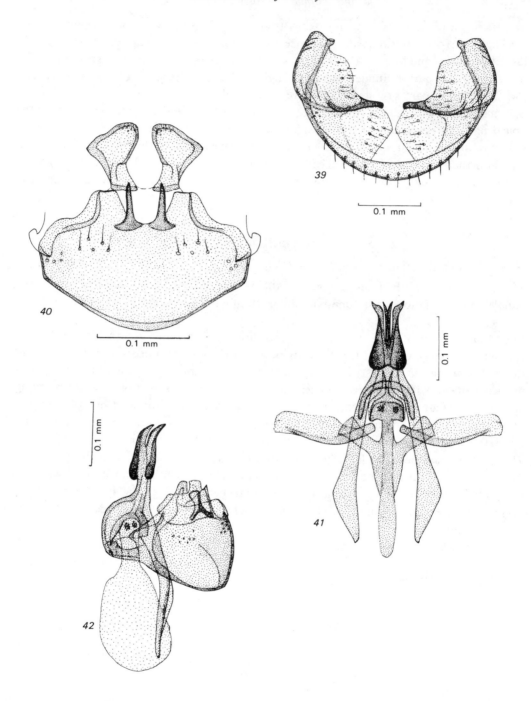

Figs. 39–42: *Glabellula* sp. no. 1
39. epandrium; 40. gonopods; 41. aedeagus;
42. genitalia, lateral

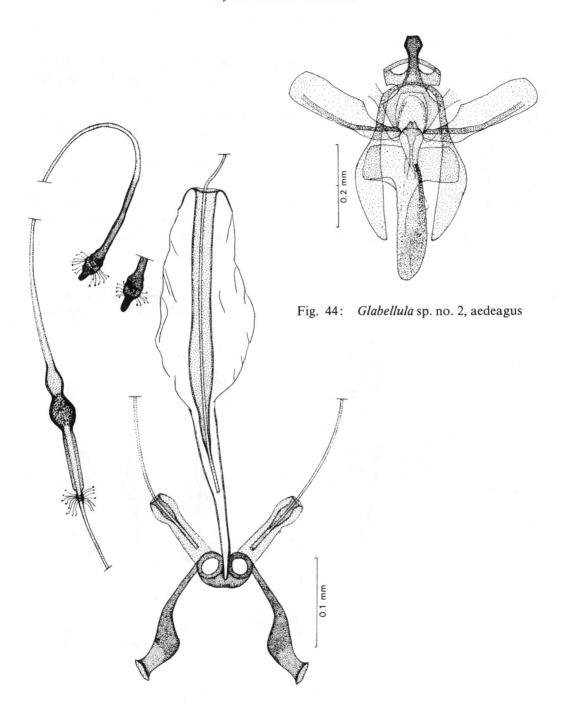

Fig. 44: *Glabellula* sp. no. 2, aedeagus

Fig. 43: *Glabellula* sp. no. 1, spermatheca

Doliopteryx Hesse, 1956

One new species (Figs. 45, 46)

This genus was known until now only from two species from South West Africa, but the new species was found to be widely distributed in the southern Negev and the Arava. It is probably more widely distributed in Africa but has been overlooked.

Epandrium short, rounded, with two lateral posterior processes. Gonocoxites small, rounded. Dististyli short, wide, with rounded apex. Aedeagus conical; sheath with two long, posterior processes and two short, apical processes. Apodeme large, with two short, curved, lateral processes and two rounded areas of sensilla before them.

Spermathecae with a funnel and reservoir similar to those described for *Glabellula*, but all three spermathecae are of equal size. The sperm capsules form a tapering spiral with rounded apex. Funnel angularly widened in the middle. The spermathecae open together in a sclerotized part in the vagina. Furca U-shaped, with lateral indentations.

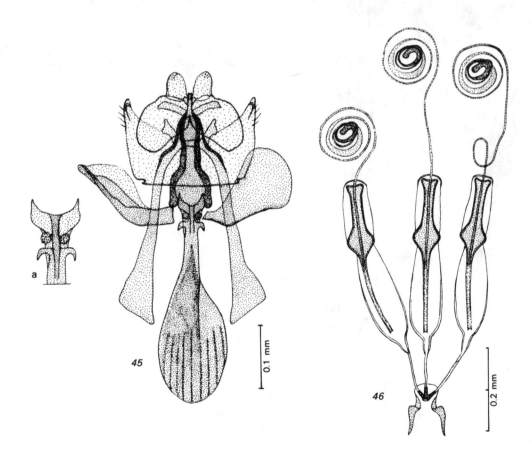

Figs. 45–46: *Doliopteryx* n. sp.
45. genitalia; (a) head of apodeme, enlarged; 46. spermathecae

New Genus
(Figs. 47 – 49)

This genus is related to *Glabellula* in having R_2 ending in R_1, forming a triangular marginal cell which is, however, much longer than in *Glabellula* and resembles this cell in *Mythicomyia* and in the Ethiopian genus *Aetheoptilus*; however, the latter two genera have a discal cell which is absent in the new genus.

The only collected male was not dissected, but the genitalia are everted and very characteristic. The aedeagus is long, narrow, tapering, curved downwards, with a slightly widened apex. Above it are two long, narrow, upcurved, black processes which are bifid at the apex. Lateral to their base are two additional black processes which are curved, forming a nearly complete ring, and appear to be attached to the gonopods.

There are only two functional spermathecae. The lateral spermathecae have large, pear-shaped, recurved capsules, continuing in a narrow duct into a curved ejection apparatus with widened ends. The median spermatheca is rudimentary, and only the basal part of the ejection apparatus is present (for details, see Fig. 49). Furca U-shaped, divided into a number of branches posteriorly.

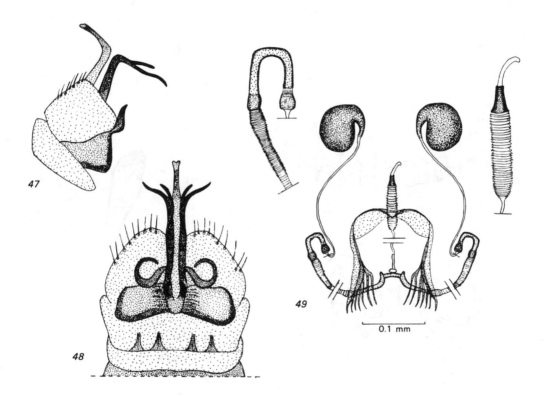

Figs. 47–49: New genus
47. male genitalia, external view, lateral; 48. same, dorsal view; 49. spermathecae

Mythicomyia Coquillett, 1893

M. armata, atra, oporina, rileyi, triformis (Figs. 50–57)

The male genitalia are very small and their parts superimposed, so that it is sometimes difficult to identify them.

Epandrium undivided, rounded, with posterior, lateral processes of specific form. Gonocoxites small, of irregular form, with processes of varying form. Dististyli long, triangular, with long, narrow, slightly curved apex in *triformis*, shorter in the other species. Aedeagus conical, narrow, with broad, bulging, basal part. The apical part is sometimes situated at a right angle to the base so that not all parts are recognizable in the same

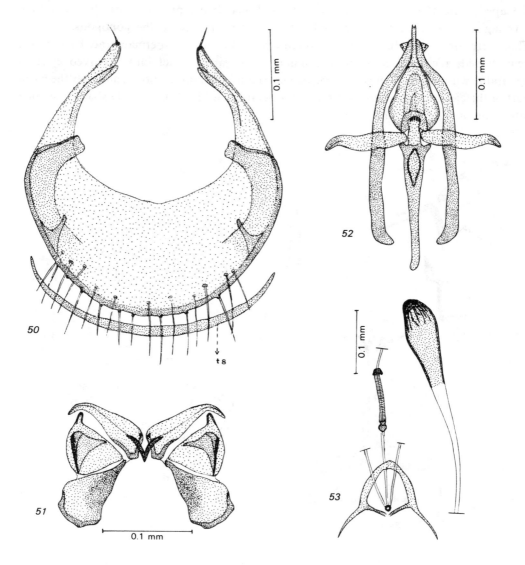

Figs. 50–53: *Mythicomyia triformis*
50. epandrium; 51. gonopods and dististyli; 52. aedeagus; 53. spermatheca

preparation. Sheath with two long, posterior processes. Apodeme with two short processes near its head in some species. There are apical processes of specific form in *oporina*, the connection of which could not be made out (Fig. 54). The drawings of the male genitalia in lateral view given by Hull (1973) and Hall (1975) are partly incorrect. The part labelled 'aedeagus' by Hall is in fact the narrow tergite 8 and so is the part labelled 'epiphallus' by Hull (Fig. 884).

Spermathecae with capsules of varying form (Figs. 53, 55–57) and long ducts which are at first wider and then become very thin. Ejection apparatus rod-like, striated, with a rounded or flattened apical end plate and also a basal end plate in some species. The ducts end in a sclerotized part in the vagina. Furca with a wide, U-shaped anterior part having two long inner processes and two long, diverging posterior processes.

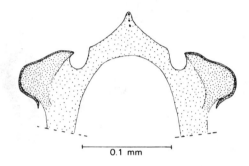

0.1 mm

Fig. 54: *Mythicomyia oporina*, apical processes of male genitalia

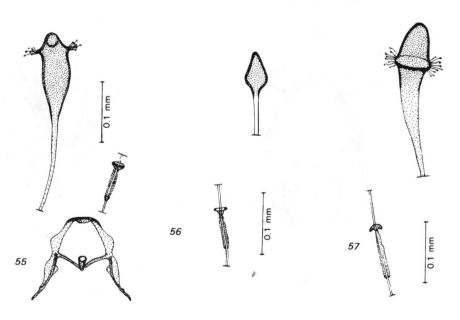

Figs. 55–57: Spermatheca
55. *Mythicomyia atra*; 56. *Mythicomyia oporina*; 57. *Mythicomyia rileyi*

31

Platypygus Loew, 1844

P. chrysanthemi, kurdorum, ridibundus (Figs. 58–62)

Epandrium short, with concave posterior margin and two dark, posterior, lateral processes. Gonocoxites short, rounded; dististyli broad, of specific form, bifid at the apex in *ridibundus*. Aedeagus very large, its greater part situated proximal to the gonopods. The wide basal part bears a separate plate at the base, below which ends the relatively small apodeme which has two broad, flattened lateral processes near its apex. The sheath extends to near the end of the narrow part of the aedeagus. It has two long proximal and two lateral processes.

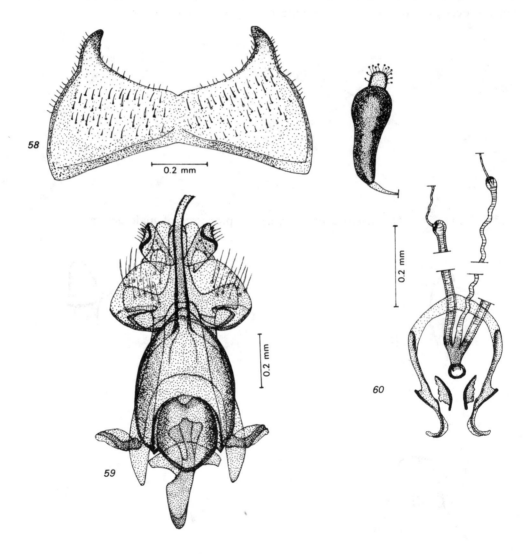

Figs. 58–60: *Platypygus chrysanthemi*
58. epandrium; 59. gonopods and aedeagus; 60. spermatheca

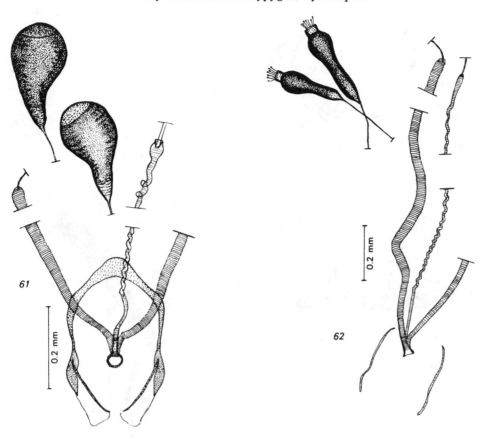

Fig. 61: *Platypygus kurdorum*, spermathecae
Fig. 62: *Platypygus ridibundus*, spermathecae

The spermathecae form pear-shaped capsules of specific form. They are moderately wide in *chrysanthemi*, with a small, rounded apical process on which the gland is situated, narrower in *ridibundus*, with a small, cylindrical apical process, and very wide, without an apical process, in *kurdorum*. The ducts are at first very thin and then enter a long, wide striated part in the two lateral spermathecae; however, in the median spermatheca this part is much narrower and spirally twisted. Furca broadly U-shaped, narrowing posteriorly in some species, and with specific differentiations, with two separate, thin bars in *ridibundus*.

Cyrtisiopsis Séguy, 1930

C. crassirostris, melleus and a new species (Figs. 63–73)

In the past *C. melleus* was placed in the genus *Platypygus* which it resembles in wing venation. However, it differs so markedly from the other species of *Platypygus* in its long proboscis, the prolonged occiput and other characters that Séguy established for it the

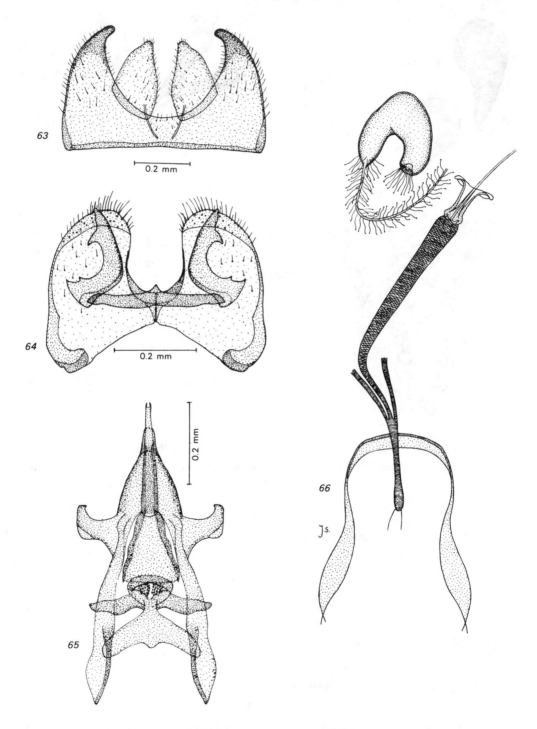

Figs. 63–66: *Cyrtisiopsis melleus*
63. epandrium; 64. gonopods; 65. aedeagus; 66 spermatheca

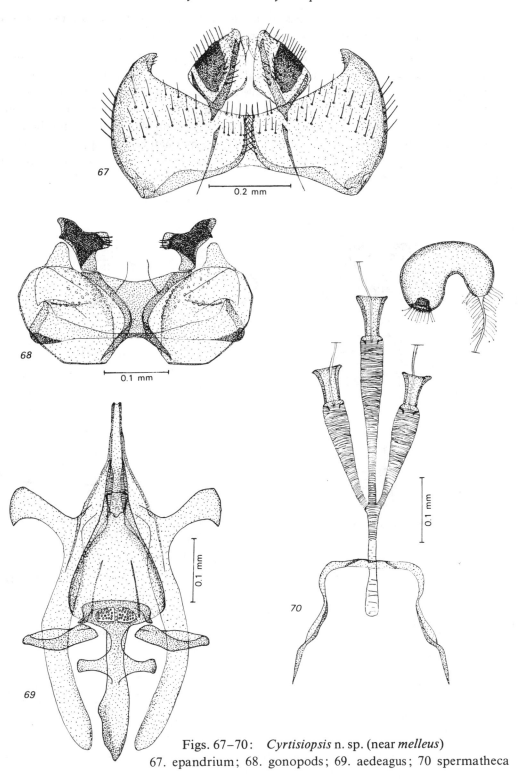

Figs. 67–70: *Cyrtisiopsis* n. sp. (near *melleus*)
67. epandrium; 68. gonopods; 69. aedeagus; 70 spermatheca

genus *Cyrtisiopsis* with the type species *singularis* which Engel made a synonym of *melleus* Loew, 1856. This thus becomes the type species of the genus.

C. melleus. Epandrium short, deeply concave posteriorly, with two posterior lateral processes. Gonocoxites broad, rounded apically. Dististyli long, narrow, curved, with two pointed apical processes situated in the middle of the gonocoxites. Aedeagus conical, with slightly curved apex. Sheath extending to near its apex, with two lateral and two long, proximal processes. Apodeme short and wide, nearly triangular, with a wide head. Spermathecae with oblong, curved capsules. All three spermathecae of equal length. Ducts thin. Ejection apparatus very large, with cup-shaped apical part, striated, narrowing proximally and forming a relatively long common duct. Furca U-shaped.

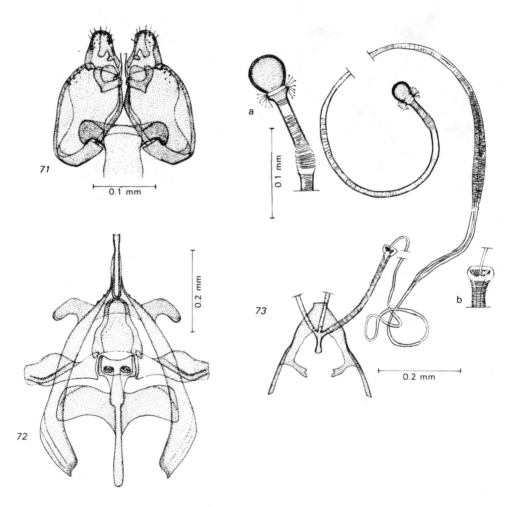

Figs. 71–73: *Cyrtisiopsis crassirostris*
71. gonopods; 72. aedeagus; 73. spermatheca;
(a) capsule enlarged; (b) apex of ejection apparatus

A new species, closely resembling *melleus* in habitus, has a rounded head, not prolonged posteriorly as in *melleus*, and differs also in the genitalia and other characters. The aedeagus is much wider; the dististyli are short, irregularly triangular and situated at the apex of the gonocoxites. The cerci are triangular and have a dark area in the middle. The spermathecae resemble those of *melleus* but the median spermatheca is distinctly longer than the lateral ones.

The prolonged occiput of *melleus* was considered to be a generic character. However, the new species has a rounded head, not prolonged posteriorly, but the ventral sulcus of the head is the same as in *melleus*. Efflatoun apparently had both species before him and a specimen determined by him as *melleus* belongs to the new species. His drawings (1945) of the head of *melleus* are also of the new species.

C. crassirostris. The male genitalia resemble those of *melleus* but the gonocoxites are normal. Dististyli situated apically, broadly triangular, with rounded apex. The lateral processes of the apodeme of the aedeagus are very wide.

The spermathecae are quite distinct. Capsules very small, globular, continuing in a wide, sclerotized duct which widens in the middle and then again becomes narrow and transparent. Ejection apparatus striated, with cup-shaped apex. Furca V-shaped, with two inner processes.

Subgenus **Ceratolaemus** Hesse, 1938

C. xanthogramma (Figs. 74–77)

This species closely resembles *Cyrtisiopsis melleus* in habitus, the prolonged occiput, the form of the antennae and other characters. It differs from *melleus* mainly in the absence of a discal cell. An illustration of the male genitalia was given by Hesse (1938).

Epandrium very short, concave posteriorly, with long, narrow, posterior, lateral processes. Cerci broadly triangular. Gonocoxites triangular; dististyli triangular with a pointed apical process, situated at the apex of the gonocoxites. Aedeagus broadly conical; sheath with two curved apical processes. This differs in some characters from the illustration given by Hesse and the specimen examined (from Tanganyika) may belong to another species.

Spermathecae (paratype of *xanthogramma*) with large, ovoid, transversely striated capsules. Ducts thin (partly lost). Ejection apparatus short, striated, with wide, cup-shaped apex in which an oblong, membranous part is situated. The ducts end together in a sclerotized part in the vagina. Furca broadly U-shaped, with two narrow inner processes. The ejection apparatus closely resembles that of *Cyrtisiopsis melleus*.

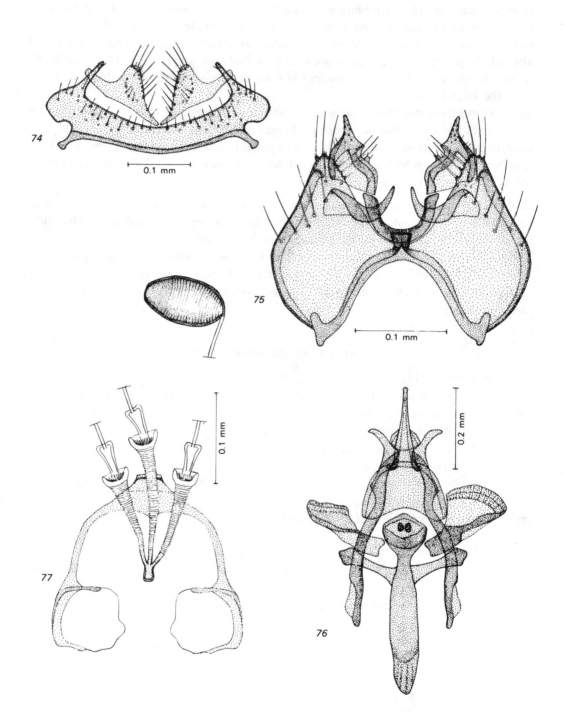

Figs. 74–77 : *Cyrtisiopsis (Ceratolaemus) xanthogramma*
74. epandrium ; 75 gonopods ; 76. aedeagus ; 77. spermatheca

GERONINAE Hesse, 1938

Geron Meigen, 1820

G. (?) gibbosus, intonsus, longiventris, mystacinus and numerous undescribed species (Figs. 78 – 118)

The species are very similar and difficult to distinguish on the basis of external characters. However, the genitalia of both sexes are highly differentiated and different from those of

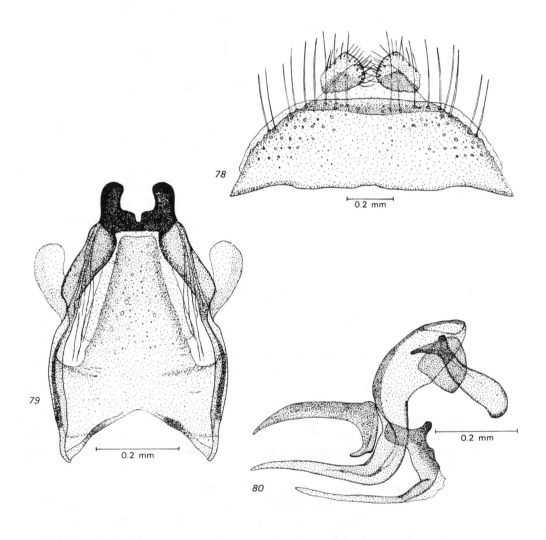

Figs. 78–80: *Geron (?) gibbosus*
78. epandrium; 79. gonopods; 80. aedeagus

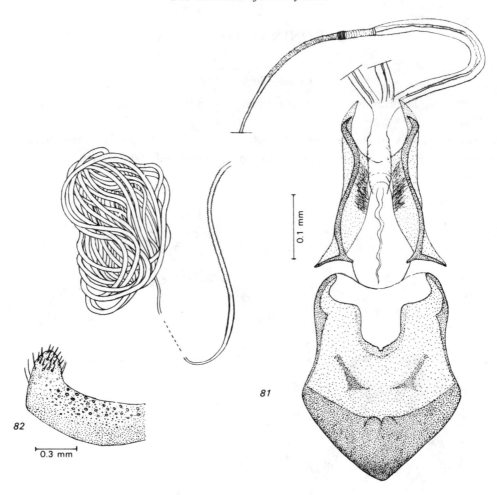

Figs. 81–82:   *Geron* (?) *gibbosus*
81. spermatheca (basal part) and capsule; 82. tergite 8 of female

the other Bombyliidae. Hesse (1938) described over 20 species from South Africa, which were previously considered as 'forms' of *gibbosus* by Bezzi and Loew. Austen (1937) recorded only two species from Palestine (one of them wrongly identified by Engel). Engel (1933) recorded only three Palaearctic species. Examination of the genitalia of a large material from Israel proved that there are over 20 species.

The identity of the type species is doubtful. The type is a single female, kept in the Paris Museum, from Beaucaire in Southern France. However, at least one other species has been recorded from Southern France by Bowden (1974). Only examination of the genitalia of the type would permit definition of its identity.

The male genitalia are complicated and highly differentiated. The epandrium is usually short, trapezoidal or rounded posteriorly, and shows only minor specific differences. The

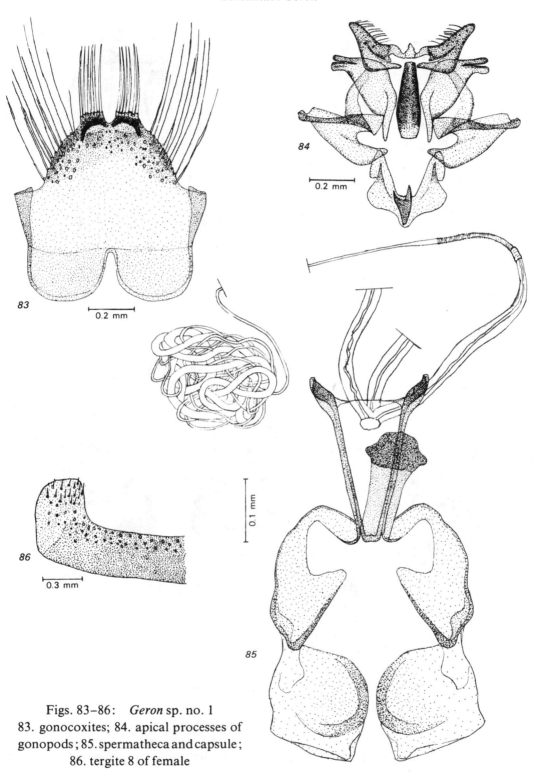

Figs. 83–86: *Geron* sp. no. 1
83. gonocoxites; 84. apical processes of
gonopods; 85. spermatheca and capsule;
86. tergite 8 of female

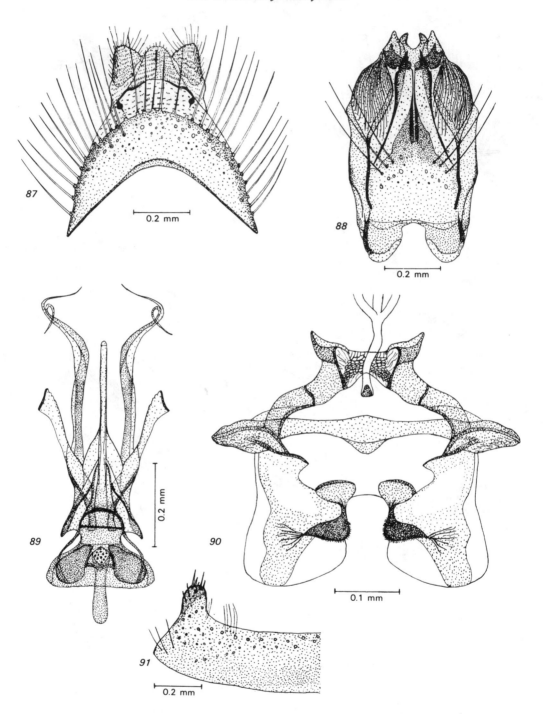

Figs. 87–91: *Geron mystacinus*
87. epandrium; 88. gonopods; 89. aedeagus;
90. furca and adjacent sclerites of female; 91. tergite 8 of female

gonocoxites are fused and bear at the apex a movable part which can be folded back and has a number of processes of varying form, one pair of which is probably homologous to the dististyli of other Bombyliidae. In two of the species examined (*longiventris* and an American species near *snowi*), one pair of processes is articulated, larger than the others and closely resembles dististyli. In some species, the gonocoxites bear one or two rows of long spines at the apex, which are clearly visible externally. The aedeagus is of specific form in most species examined; it may be relatively short and wide or long and narrow. Its basal part is widened and the small apodeme is usually situated at a right angle to the long axis of the aedeagus. The processes of the sheath vary greatly in form and length. They may be simple, very long, bearing a brush of hairs at the end, or twisted and divided at the end (*mystacinus*).

The female genitalia have proved of particular importance. The capsules of the spermathecae form a dense coil of narrow tubes in all species examined. The ducts are very long and thin and bear one or two small sclerotizations of varying form in their proximal

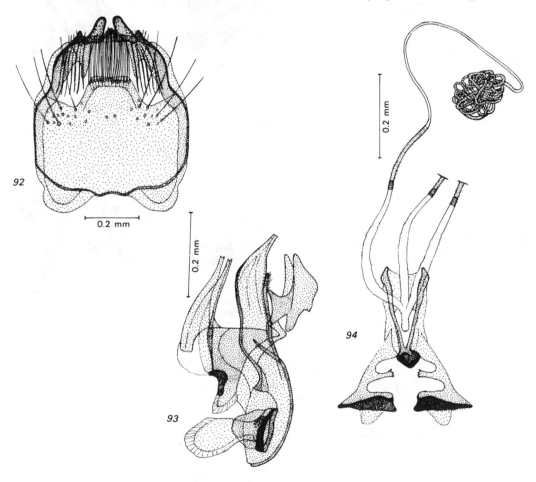

Figs. 92–94: *Geron* sp. no. 2
92. gonopods; 93. aedeagus; 94. spermatheca

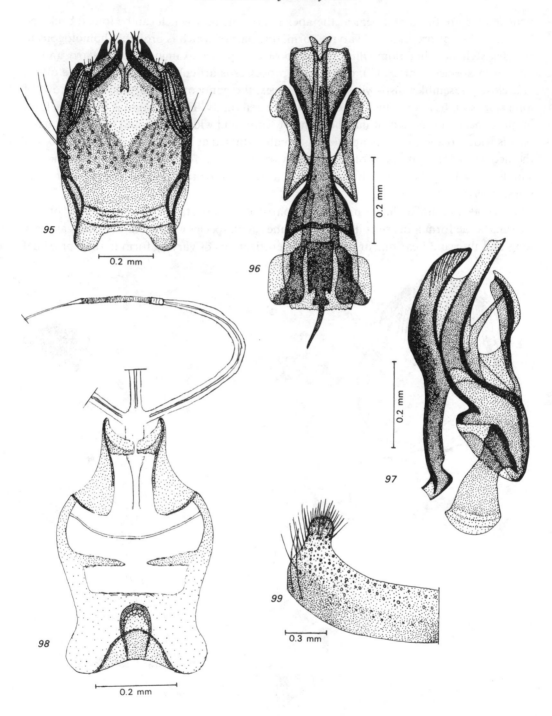

Figs. 95–99: *Geron intonsus*
95. gonopods; 96. aedeagus; 97. same, lateral;
98. furca of female; 99. tergite 8 of female

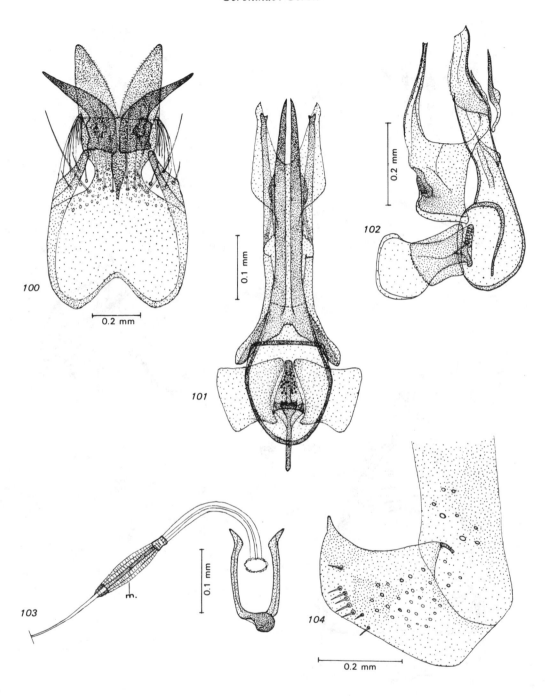

Figs. 100–104: *Geron longiventris*
100. gonopods; 101. aedeagus; 102. same lateral;
103. furca of female; 104. tergite 8 of female

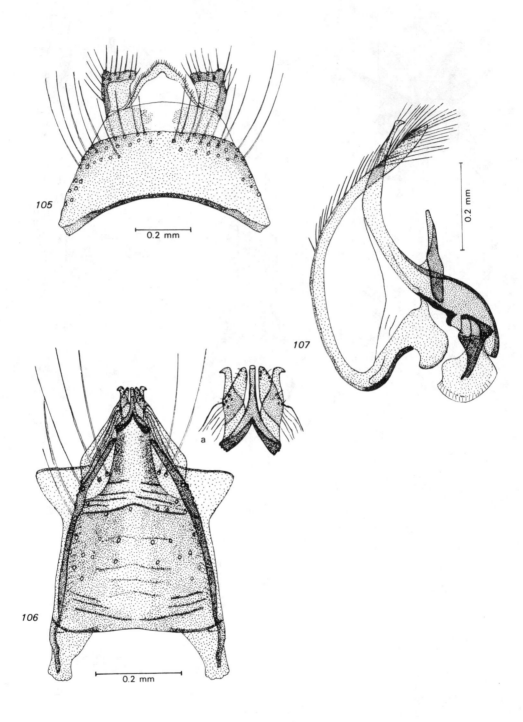

Figs. 105–107: *Geron* sp. no. 3
105. epandrium; 106. gonopods; (a) apex enlarged; 107. aedeagus

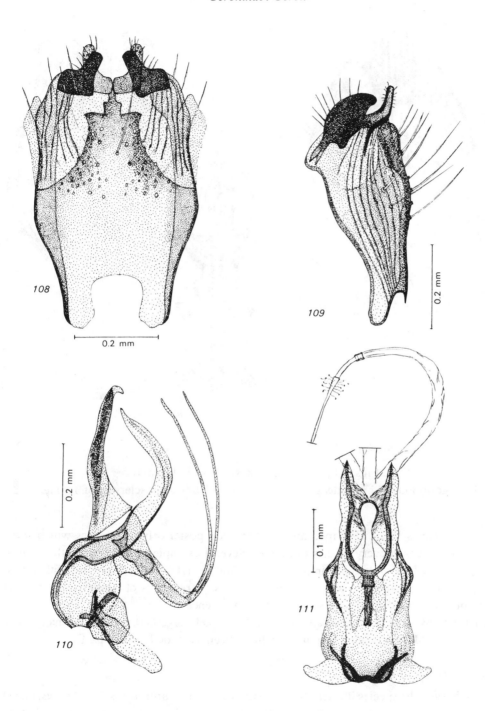

Figs. 108–111: *Geron* sp. no. 4
108. gonopods; 109. same, lateral;
110. aedeagus; 111. furca and adjacent sclerites

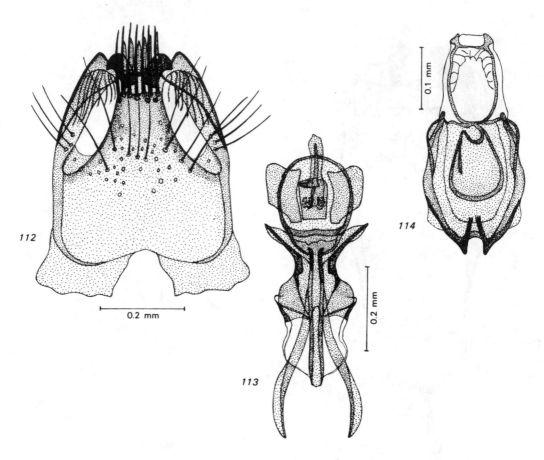

Figs. 112–114: *Geron* sp. no. 5 (Egypt)
112. gonopods; 113. aedeagus; 114. furca and adjacent sclerites of female

part. The furca is V- or U-shaped and is connected posteriorly to sclerites which show specific differences in every species examined. Several examples are shown in the figures. This has proved of particular importance as the females are even more difficult to determine by external characters than are the males. Tergite 8 is of similar form in most species but its lateral processes show specific differences.

The surprisingly large number of species found in Israel suggests that there are many more species in the Mediterranean region than have been recorded in the past.

Notes on Species (Figs. 78–82)

G. (?) *gibbosus*. The species illustrated is considered by most authors as *G. gibbosus*, based mainly on specimens from the Eastern Mediterranean. Its occurrence in the Western Mediterranean is not certain, as another species (*G. corcyreus* Frey) is common in Southern France according to Bowden (1974). The identity of *G. gibbosus* is doubtful, as stated above, and will have to be determined by the examination of the genitalia of the type.

Species no. 1. This species occurs together with *G.* (?) *gibbosus* and resembles it closely, but differs distinctly from it by the presence of two groups of setae on the gonocoxites and in the genitalia in both sexes (Figs. 83–86).

G. mystacinus. The male is easily recognized by the black face beard. The female has not been described. It has been identified by the collection of series in isolated localities in which it was associated only with males of *mystacinus*. The female differs from the male in its white face beard and was in the past wrongly identified as that of *G. krymensis* or *intonsus*. The aedeagus is very long and the two processes of the sheath are much longer than the aedeagus, twisted and divided into two long, thin processes at the end. The furca and the adjacent sclerites are shown in Fig. 90.

Species no. 2. This species was wrongly identified as *krymensis* or *intonsus* (Engel, 1933). The gonocoxites bear a dense row of long setae at the apex. The spermathecae are shown in Fig. 94.

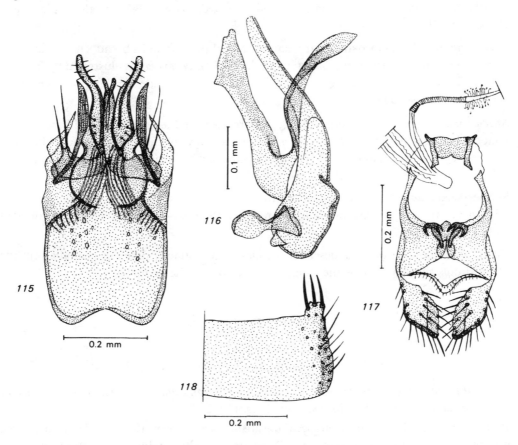

Figs. 115–118: *Geron* sp. (near *snowi*)
115. gonopods; 116. aedeagus; 117 furca and
adjacent sclerites of female; 118. tergite 8 of female

G. intonsus. This species is easily identified by the wide frons in both sexes. It is known so far only from the types collected in Egypt and a long series of specimens from the southern Negev. The male genitalia are very characteristic. The aedeagus is strongly sclerotized, black, with a relatively wide apical opening. The aedeagal process has a wide, nearly square median plate and two narrower, shorter lateral processes. Furca shown in Fig. 98. Records of *intonsus* from Northern Israel and Lebanon refer to species no. 2, and records from Southern Russia probably to *krymensis*.

G. longiventris. The male is easily recognized by two pairs of externally visible processes of the gonocoxites, one pair of which is probably homologous to the dististyli. Aedeagus long and narrow; sheath with four long apical processes. The furca is V-shaped with a thick sclerotization at the base (Fig. 103).

Species no. 3. Only a single male was collected. This is a very small, shining black species resembling some South African species (*nigerrimus* and others). The processes of the sheath are very long and narrow and covered with setae in the apical half (Figs. 105 – 107).

Species no. 4. The gonocoxites bear a narrow apical process and a broad, curved, laterally compressed process. The genitalia resemble those of *G. corcyreus* as illustrated by Bowden (1974, Fig. 7, p.96). The aedeagus is long and curved, and so are the thin processes of the sheath (Figs. 108 – 111).

Species no. 5. Two specimens from Egypt were identified as *gibbosus* by Efflatoun. The gonocoxites bear two thick apical processes and two rows of short, thick setae at the apex. The aedeagus is shown in Fig. 113. The furca of the female is shown in Fig. 114. The female belongs to another species which also occurs in Israel.

Species near *snowi* (Nearctic). Two of the apical processes of the gonocoxites closely resemble dististyli. They are long, narrow, pointed, slightly S-curved, tapering to a rounded end (Figs. 115 – 118).

There are about ten other undescribed species with distinctly different genitalia, but the illustrations give an idea of the range of variation of the genitalia.

USIINAE Becker, 1912

Usia Latreille, 1802

U. aurata, bicolor, (?) *carmelitensis, florea, forcipata, ignorata, lutescens, ornata,* (?) *pusilla versicolor* and several undescribed species (Figs. 119–180)

Past attempts to divide this genus seem unsatisfactory. Becker (1906) divided it into three groups according to colour characters. Paramonov (1929) established the subgenus *Parageron* for species with holoptic males, but this is also apparently incorrect as holoptic males are present in quite different species groups. One species with holoptic males is closely related to *Usia aurata*. There is apparently only one group of species which is well defined by external characters and may have to be considered as a subgenus, the small black species

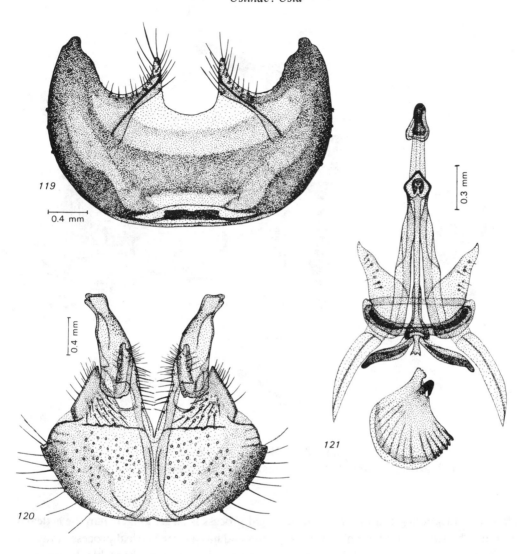

Figs. 119–121: *Usia florea*
119. epandrium; 120 gonopods; 121. aedeagus, dorsal

in which the males have a wide frons and a very large hypopygium which projects ventrally and is sometimes as large as the entire remaining part of the abdomen.

The epandrium is often of characteristic form, with posterior processes and projecting more or less ventrally. These processes are visible externally in *forcipata* and are very wide and surround the anus in (?) *carmelitensis*. The gonocoxites are also of specific form and the dististyli are distinctly different in nearly every species examined. The aedeagus is highly differentiated in most species. It is conical in most species, either straight or bent

51

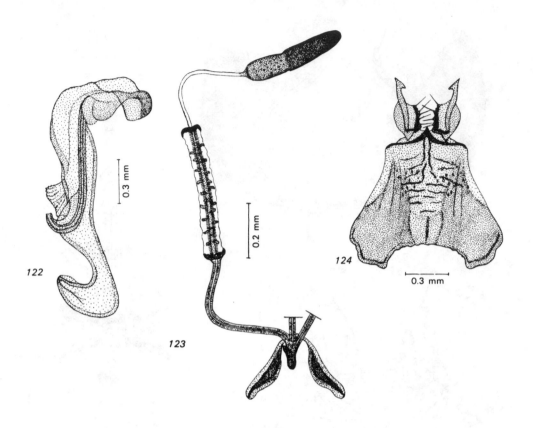

Figs. 122–124 : *Usia florea*
122. aedeagus, lateral ; 123. spermatheca ; 124. sclerite behind furca

almost at a right angle at the end. The aedeagal process is large, longer than the aedeagus, and may be simple and flattened. It is very long and has a large vertical process at the apex in the type species *florea*. In the small black species, it may bear short, black spines in a characteristic arrangement, which are sometimes situated on a special process (*forcipata, ignorata*) (Figs. 155, 159).

The spermathecae are of similar form in most species, but with specific differences. The capsules are sausage- or pear-shaped, the ducts are thin, and the ejection apparatus is well developed, with small and larger processes and distinct end plates. The processes are sometimes divided into several points at the apex. Furca of characteristic form in many species, with a sclerite of specific form between the lateral arms.

U. aurata has been recorded throughout the Mediterranean region. However, comparison of the species from Israel and Egypt determined as *aurata* with specimens from Corsica, Spain and Algeria showed that the species from Israel, one of which has holoptic males, differ both in external characters and in the genitalia from *aurata* and will have to be described as new species. *U. aurata* is apparently mainly a West Mediterranean species but

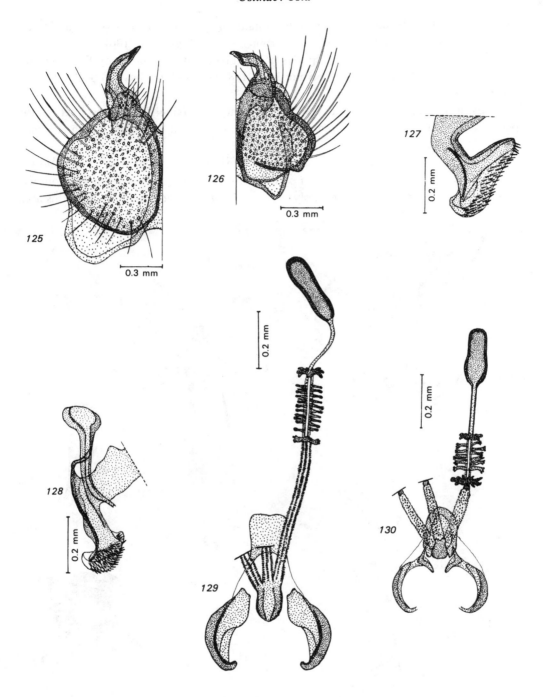

Figs. 125–130: *Usia aurata*
125. gonopod (Spain); 126. same (Algeria);
127. apex of aedeagus (Algeria); 128. same (Spain);
129. spermatheca (Algeria); 130. same (Corsica)

it may also consist of local forms, as the genitalia of specimens from Corsica and Spain differ distinctly from those of specimens from Algeria (Figs. 125–130).

A small black species from Egypt (species no. 6), identified as *ignorata* by Efflatoun, has completely different genitalia, as shown in Figs. 178–180, and will have to be described as a new species.

U. forcipata was recorded by Engel (1932) from Rehovot in Israel, but no specimens of this species were found in the large material from the same area examined which consisted mainly of *ignorata*.

A small species collected in Western Galilee is provisionally considered as *carmelitensis* as no male was available for comparison. The record of *carmelitensis* from the southern coastal plain by Austen (1937) probably refers to *ignorata* which is common in this area.

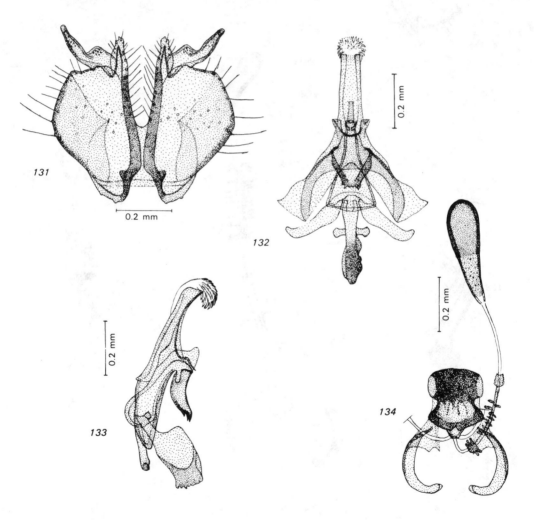

Figs. 131–134: *Usia* sp. no. 1 (near *aurata*)
131. gonopods; 132. aedeagus; 133. same, lateral; 134. spermatheca

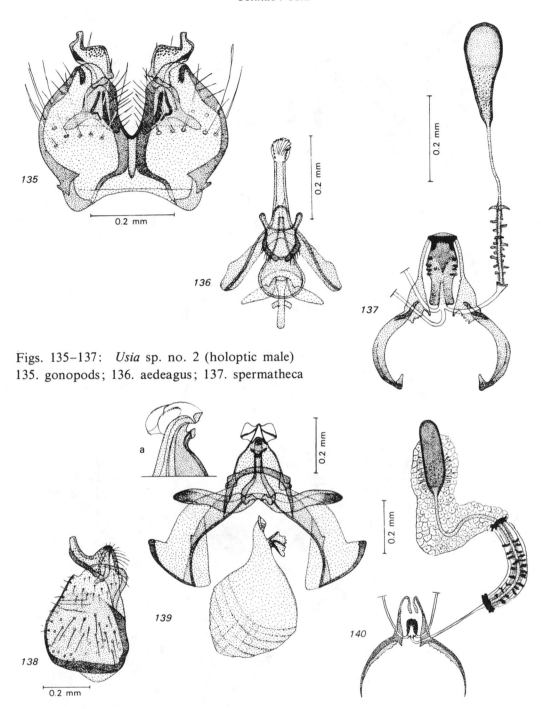

Figs. 135–137: *Usia* sp. no. 2 (holoptic male)
135. gonopods; 136. aedeagus; 137. spermatheca

Figs. 138–140: *Usia bicolor*
138. gonopod; 139. aedeagus; (a) same, apex, lateral; 140. spermatheca

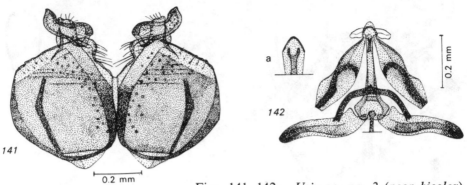

Figs. 141–142: *Usia* sp. no. 3 (near *bicolor*)
141. gonopods; 142. aedeagus; (a) same, apex enlarged

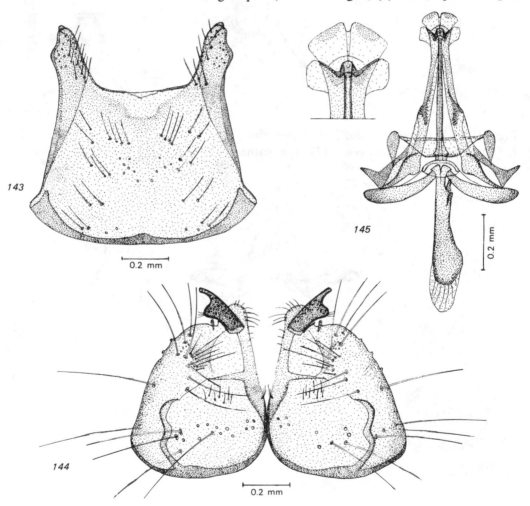

Figs. 143–145: *Usia versicolor*
143. epandrium; 144. gonopods; 145. aedeagus

56

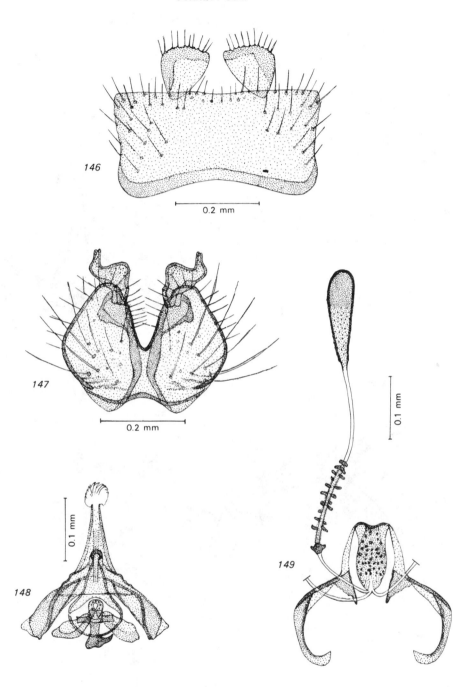

Figs. 146–149: *Usia ornata*
146. epandrium; 147. gonopods; 148. aedeagus; 149. spermatheca

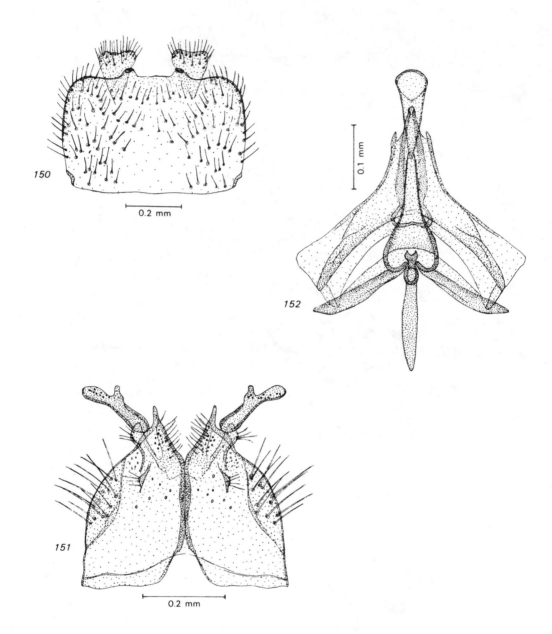

Figs. 150–152: *Usia lutescens*
150. epandrium; 151 gonopods; 152. aedeagus

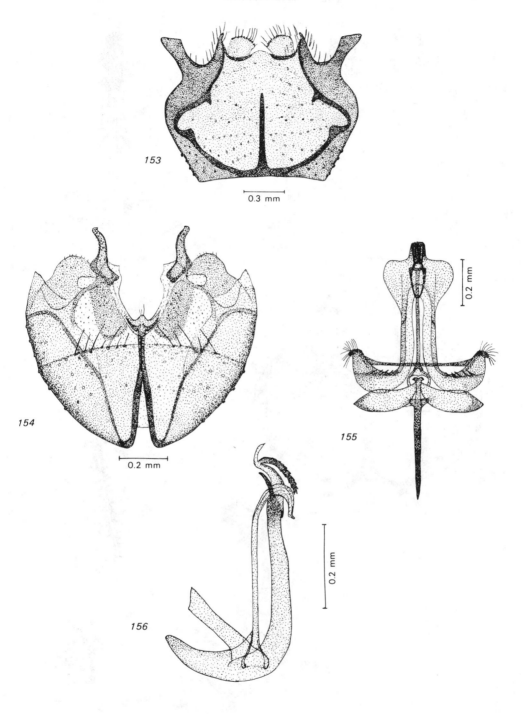

Figs. 153–156: *Usia forcipata*
153. epandrium; 154. gonopods; 155. aedeagus; 156. same, lateral

59

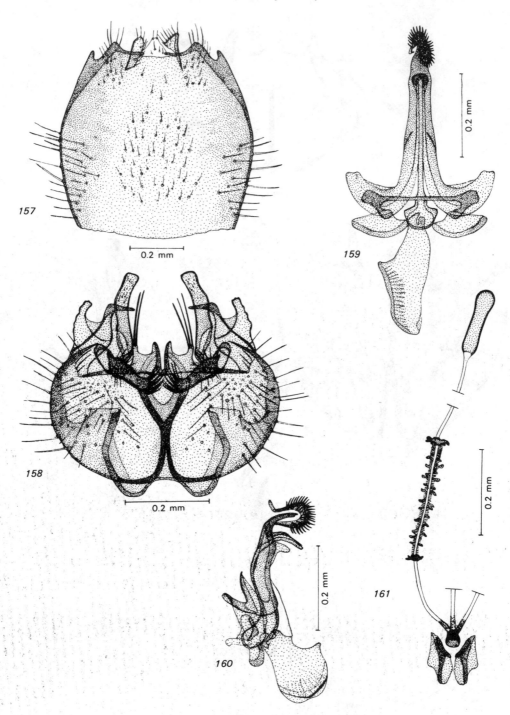

Figs. 157–161: *Usia ignorata*
157. epandrium; 158. gonopods;
159. aedeagus; 160. same, lateral; 161. spermatheca

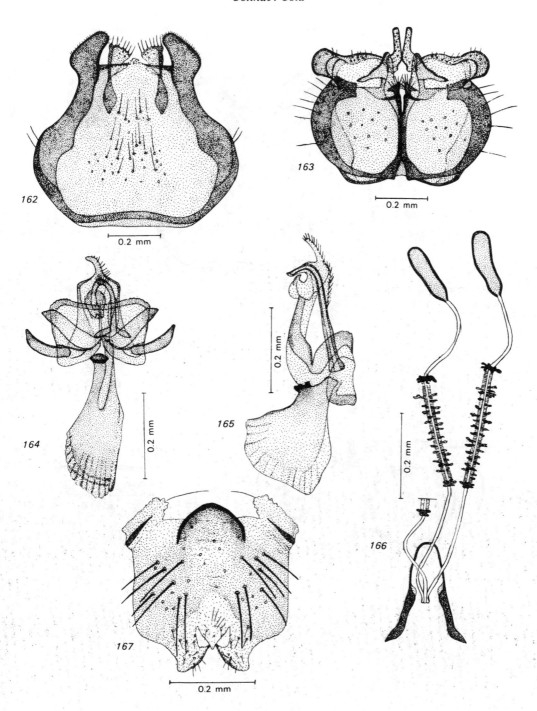

Figs. 162–167: *Usia* (?) *carmelitensis*
162. epandrium; 163. gonopods; 164. aedeagus;
165. same, lateral; 166. spermatheca; 167. sclerite behind furca

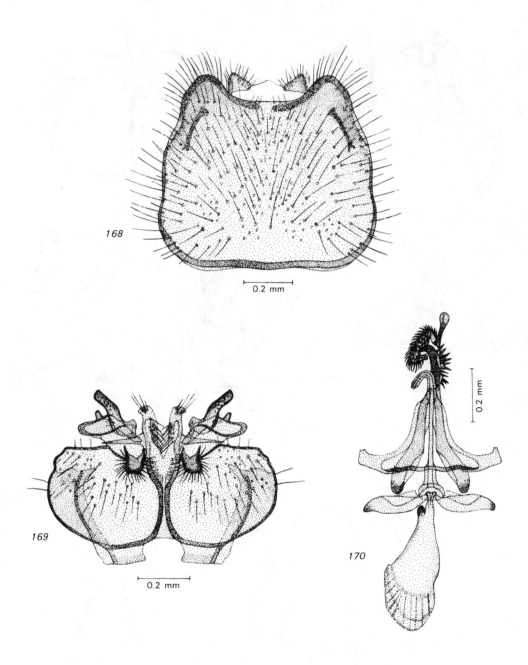

Figs. 168–170: *Usia* (?) *pusilla* (Algeria)
168. epandrium; 169. gonopods; 170. aedeagus

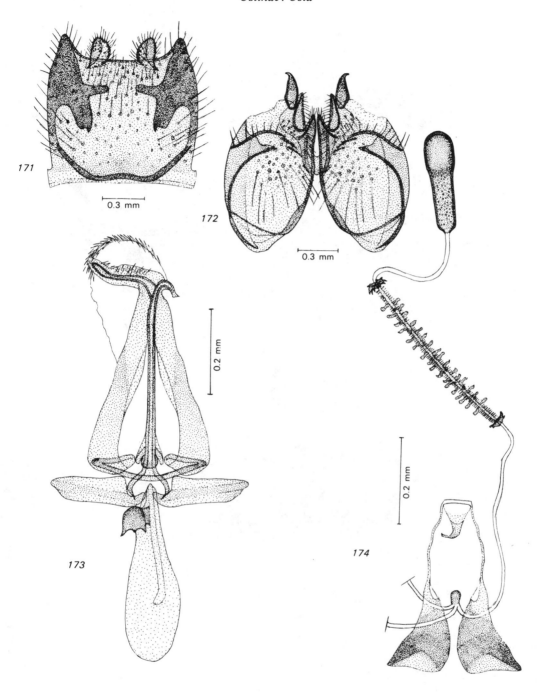

Figs. 171–174: *Usia* sp. no. 4
171. epandrium; 172. gonopods; 173. aedeagus; 174. spermatheca

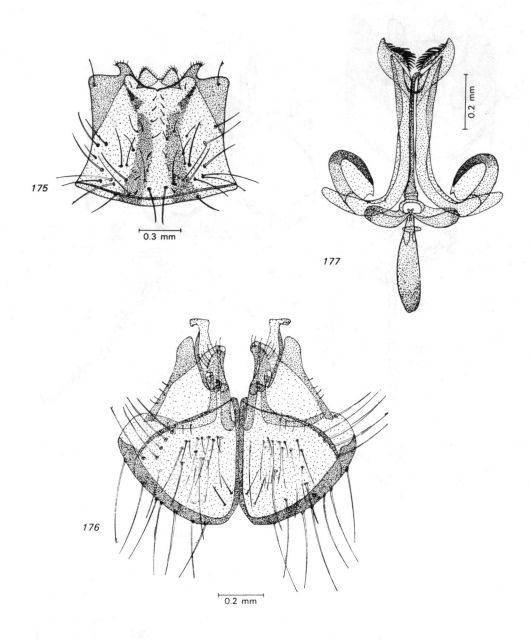

Figs. 175–177: *Usia* sp. no. 5
175. epandrium; 176. gonopods; 177. aedeagus

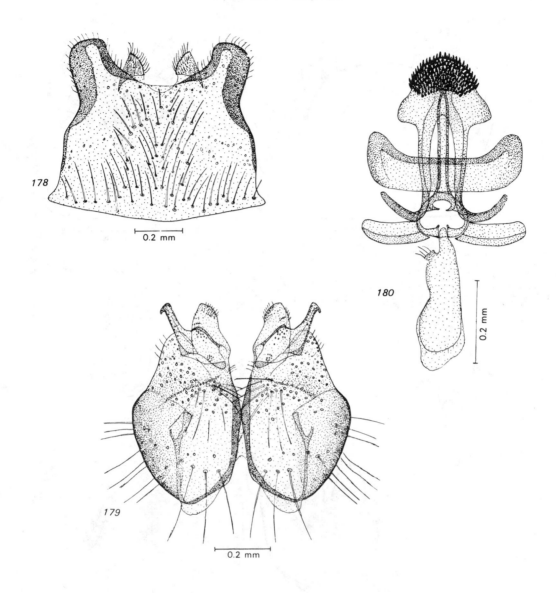

Figs. 178–180: *Usia* sp. no. 6 (Egypt)
178. epandrium; 179. gonopods; 180. aedeagus

Apolysis Loew, 1860

A. eremophila (Palaearctic), *druias* (Nearctic) (Figs. 181–186)

Epandrium rounded posteriorly. Cerci narrow, long, inserted laterally, with a small, coni-cal process in the middle. Gonocoxites partly fused, with a median process with mush-room-shaped scales at the apex. Dististyli broad, with a pointed and a rounded process

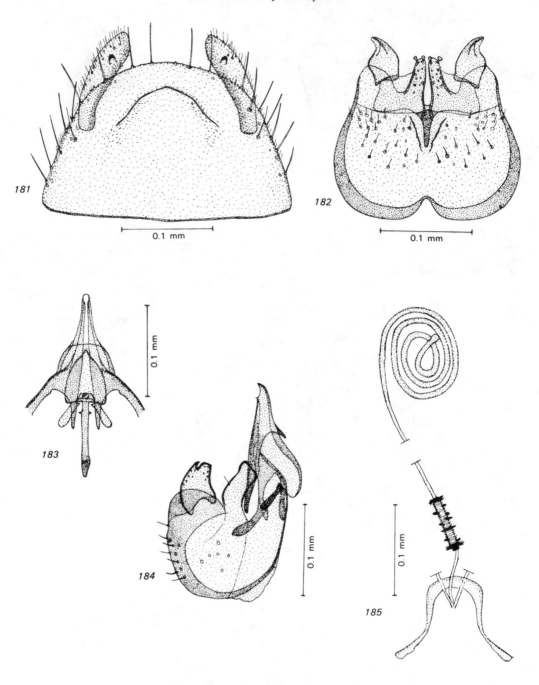

Figs. 181–185: *Apolysis eremophila*
181. epandrium; 182. gonopods; 183. aedeagus;
184. gonopods and aedeagus, lateral;
185. spermatheca

0.2 mm

Fig. 186: *Apolysis druias*, gonopods and dististylus

apically. The gonopods and dististyli of *druias* are quite different. They are broad at the base, narrowing and then widening again into a bifid apex.

The spermathecae form sclerotized spirals. The ejection apparatus is short, striated, with a few relatively large processes and end plates. Furca U-shaped, simple. The Nearctic species *druias* has similar spermathecae.

Oligodranes Loew, 1844

O. cinctura, montanus (Nearctic), and an undescribed species from Israel (Figs. 187 – 189)

New species. Epandrium nearly rectangular, with concave anterior and posterior margin and short, rounded, lateral, posterior corners. Gonocoxites broad, triangular; dististyli rounded basally, with a pointed and a rounded apical process. Aedeagus conical, simple; sheath wide, not projecting beyond apex of aedeagus.

Spermathecae with irregularly pear-shaped capsules and long, thin ducts. Ejection apparatus short, striated, lacking distinct processes, with large, cup-shaped end plates. Furca U-shaped, with two long, curved, inner posterior processes.

The spermathecae of the American species differ from those described above. Capsules of *cinctura* large, pear-shaped; ducts at first very wide, sclerotized, then narrowing. Ejection apparatus strongly sclerotized, with flower-shaped end plates and large, jagged processes. Capsules of spermathecae of *montanus* much smaller, pointed, conical, but otherwise similar. End plates of ejection apparatus small.

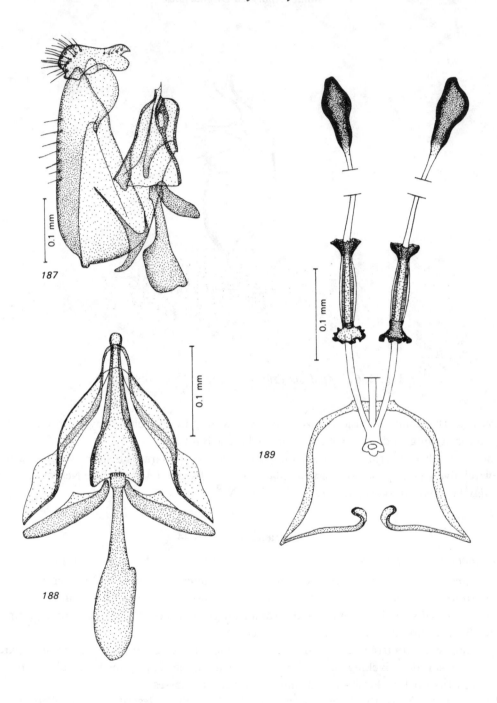

Figs. 187–189: *Oligodranes* sp.
187. gonopods and aedeagus, lateral; 188. aedeagus; 189. spermatheca

Legnotomyia Bezzi, 1902

L. bombyliiformis, trichorrhoea, and an undescribed species (Figs. 190–193)

This genus was placed by most authors in the subfamily Usiinae, but Hull (1973) included it in the Bombyliinae. It resembles both subfamilies in some characters, but differs from both in others. However, the resemblance to the Usiinae (presence of callosities on the occiput, undifferentiated abdomen of the female, absence of spines on tergite 9) seems more important than the presence of four posterior cells on the wings and the undifferentiated aedeagus.

Epandrium trapezoidal, short. Gonocoxites truncate-triangular, with long, inner, apical processes. Dististyli narrow, triangular. Aedeagus simple, conical, without differentiations and lacking an aedeagal process.

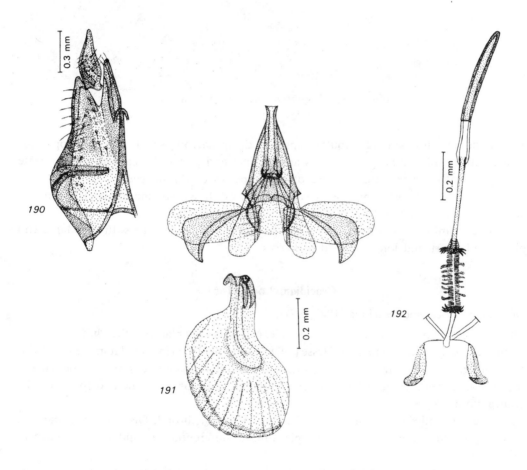

Figs. 190–192: *Legnotomyia trichorrhoea*
190. gonopod; 191. aedeagus; 192. spermatheca

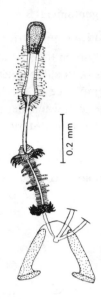

Fig. 193: *Legnotomyia* sp., spermatheca

Spermathecae with pointed, tubular capsules; ducts narrow, short. Ejection apparatus short, with small and larger processes and distinct end plates. Furca with two separate bars. Spermathecae resembling those of some species of *Anastoechus*.

One female with almost completely clear wings has different spermathecae. The capsules are much shorter than in the above two species, club-shaped and sclerotized apically, then becoming membranous, with a sclerotized base. Ejection apparatus with very large end plates and short and long, pointed processes.

Crocidium Loew, 1861

One undescribed species (Figs. 194–197)

The systematic postion of this genus is not clear. It is usually placed in the Phthiriinae, as it resembles *Phthiria* in habitus, but Hesse (1938) and Hull (1973) included it in the Bombyliinae. The genus differs from the Phthiriinae in the differentiated abdomen of the female; tergite 8 is invaginated and has a large, triangular, anterior apodeme and spines on the acanthophorites.

Epandrium rounded-rectangular. Gonocoxites triangular, broad. Dististyli long, narrow, curved, pointed. Aedeagus conical, simple, without differentiations and lacking an aedeagal process.

Spermathecae with long, pear-shaped capsules which are wider basally; ducts narrow. Ejection apparatus long, narrow, with short processes and a conical apical and rhomboidal basal sclerotization. Tergite 8 with a large triangular apodeme having a vertical median ridge. Acanthophorites short and wide, each with 4–5 long spines. Furca U-shaped, wider basally.

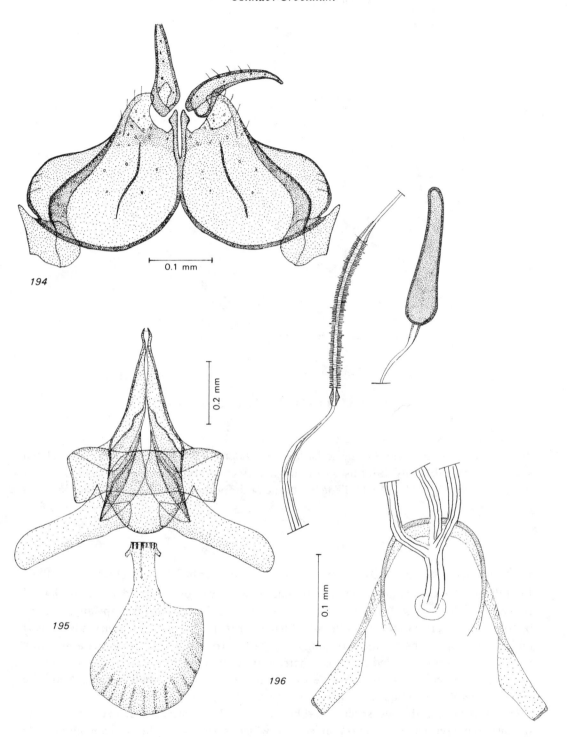

Figs. 194–196: *Crocidium* sp.
194. gonopods; 195. aedeagus; 196. spermatheca

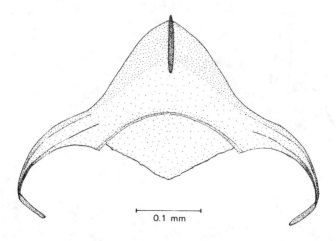

0.1 mm

Fig. 197 :   *Crocidium* sp., tergite 8 of female

HETEROTROPINAE Becker, 1912

The subfamily is considered here to contain only the genus *Heterotropus* and a new genus or subgenus for the Nearctic species *senex*. Melander (1950) placed *Prorates* and related species in this subfamily, but this is certainly incorrect. *Prorates* and related genera probably do not belong to the Bombyliidae but to the Scenopodinae, as discussed under *Prorates* (pp. 8, 18).

Heterotropus Loew, 1873

H. aegyptiacus, maculiventris, senex, and several undescribed species (Figs. 198 – 208)
Epandrium short, angularly indented proximally, with two long, narrow, posterior, lateral processes with a seta at the apex in some species, or trapezoidal, narrower apically, without processes in others. Gonocoxites broad, of irregular form. Dististyli with broad basal part and narrow apical part with a rectangularly bent part at the apex which is apparently movable. Aedeagus divided into three narrow tubes in some species, only into two such tubes in one species, and with a wide base. Lateral plates very small, curved. Apodeme large, triangular, situated at a right angle to the long axis of the aedeagus.
The gonocoxites of the new species no. 1 bear a row of long, curved setae in the middle and the dististyli have a longer, narrow, apical part which is widened at the apex and apparently not movable as in the other species. (This is the species in which the aedeagus is divided into only two prongs.)

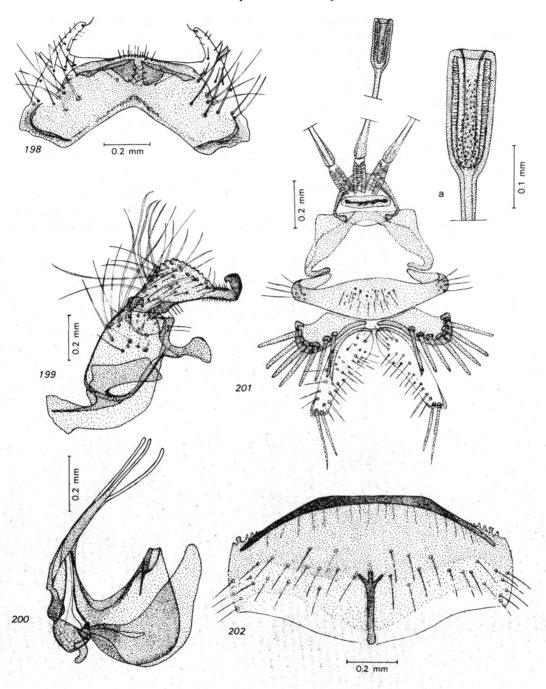

Figs. 198–202: *Heterotropus aegyptiacus*
198. epandrium; 199. gonopod; 200. aedeagus;
201. spermatheca, furca and posterior sclerites; (a) capsule enlarged;
202. tergite 8 of female

The genitalia of the Nearctic species *senex*, which is known only from the holotype, are quite different. The type was kindly sent to the author for examination by Dr L. Knutson. The epandrium is broadly rectangular, with rounded posterior corners. Gonocoxites broadly triangular, with two curved apical processes having a wider apical part. Dististyli long and narrow. Aedeagus undivided, conical, with membranous basal part. Sheath with two long, curved processes with pointed apex. Apodeme and lateral plates very small. These differences are so marked that apparently a new genus or subgenus will have to be established for *H. senex*.

The capsules of the spermathecae (Fig. 201) are broadly cylindrical, with inner ridges, a deep apical indentation and small inner spines. Ducts long, thin, ending separately in a short, sclerotized part in the vagina. Furca broadly rectangular. Acanthophorites broad, with 6 – 7 strong spines. Cerci long, with short hairs and two long setae at the apex. Tergite 8 partly invaginated, without an apodeme and with a sclerotized median ridge in its posterior part.

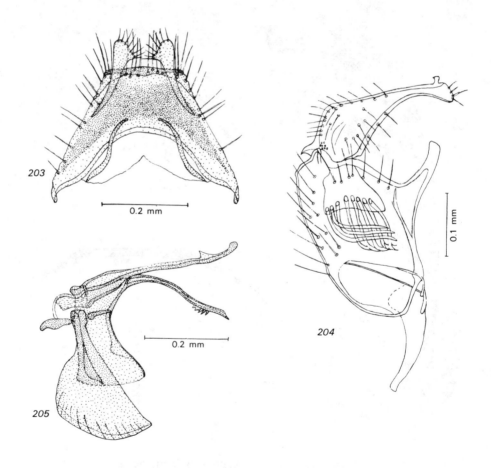

Figs. 203–205: *Heterotropus* sp. no. 1
203. epandrium; 204. gonopod; 205. aedeagus

74

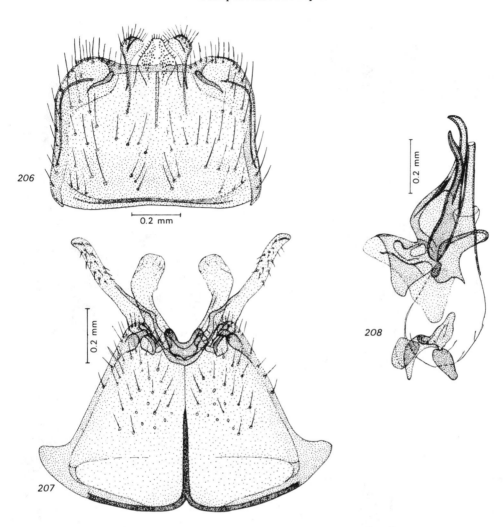

Figs. 206–208: *Heterotropus senex*
206. epandrium; 207. gonopods; 208. aedeagus

TOXOPHORINAE Schiner, 1868

The male genitalia of *Toxophora* are highly differentiated, while those of the other genera placed in this subfamily are relatively simple, but the genitalia in both sexes show marked differences. Hull (1973) placed these genera in the tribe Lepidophorini.

Toxophora Meigen, 1803

T. aegyptiaca, epargyra, maculata (Palaearctic); *virgata* (Nearctic); and three undetermined species, one of them Ethiopian (Figs. 209–235)

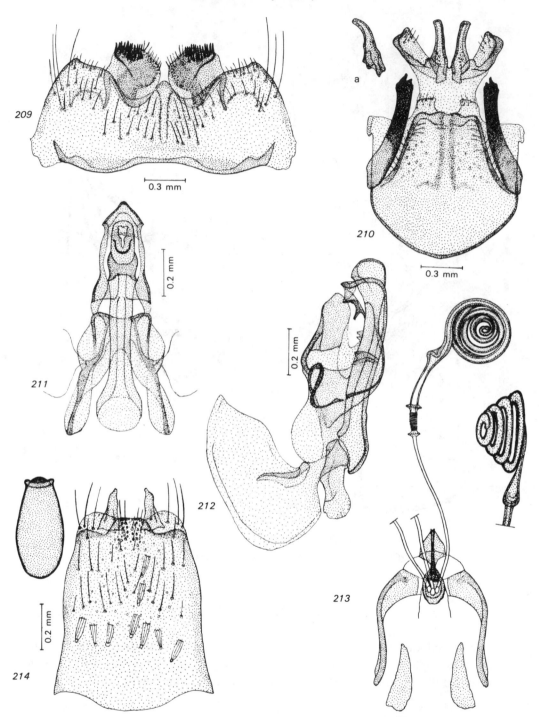

Figs. 209–214: *Toxophora maculata*
209. epandrium; 210. gonopods; (a) dististylus; 211. aedeagus;
212. same, lateral; 213. spermatheca; 214. sclerite behind furca and egg

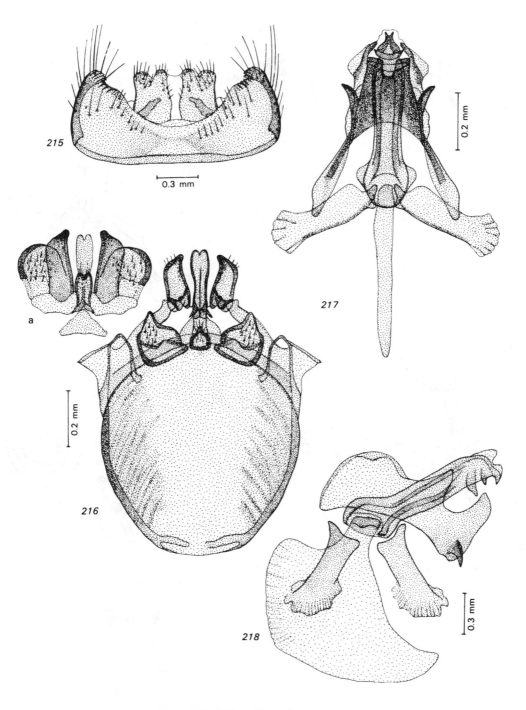

Figs. 215–218: *Toxophora epargyra*
215. epandrium; 216. gonopods; (a) apical processes, different aspect;
217. aedeagus; 218. same, lateral

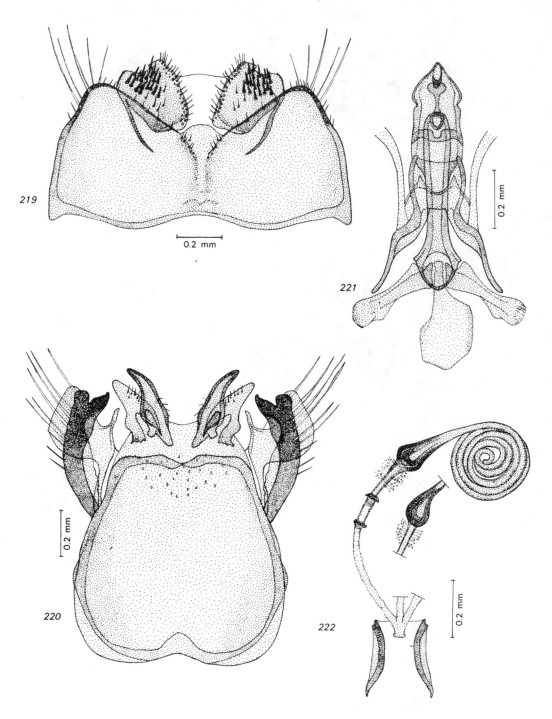

Figs. 219–222: *Toxophora virgata*
219. epandrium; 220. gonopods; 221. aedeagus; 222. spermatheca

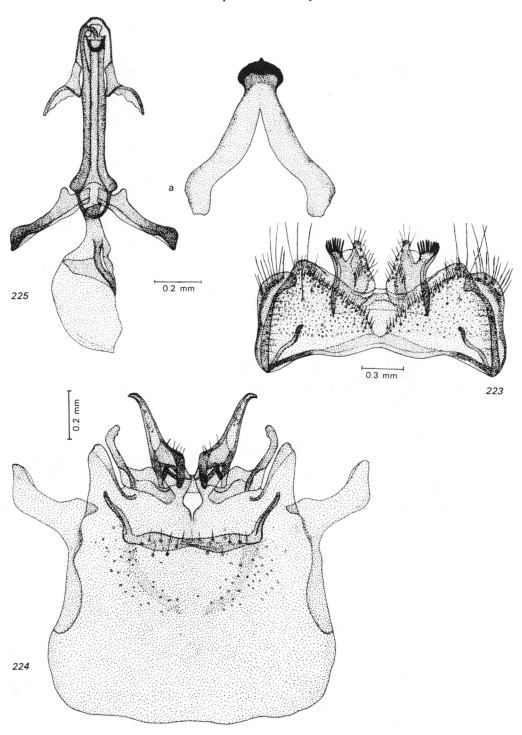

Figs. 223–225: *Toxophora aegyptiaca*
223. epandrium; 224. gonopods.; 225. aedeagus; (a) aedeagal process

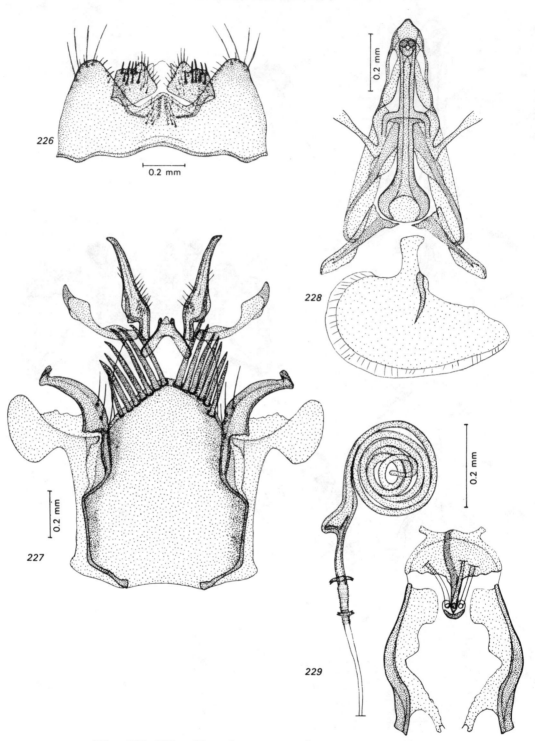

Figs. 226–229: *Toxophora* sp. no. 1
226. epandrium; 227. gonopods; 228. aedeagus; 229. spermatheca

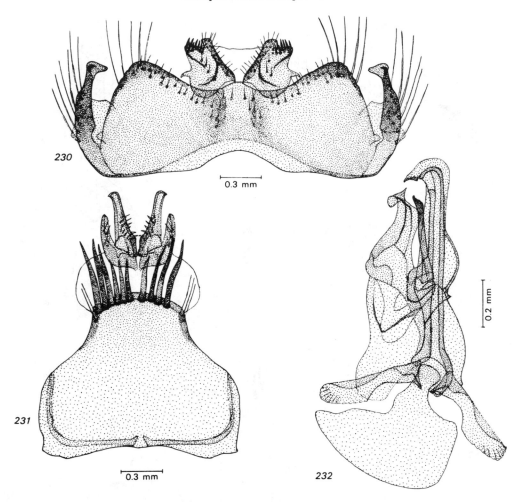

0.3 mm

0.2 mm

0.3 mm

230
231
232

Figs. 230–232: *Toxophora* sp. no. 2
230. epandrium; 231. gonopods; 232. aedeagus

Epandrium short, concave or angularly indented posteriorly. Cerci with numerous short spines in some species, bilobed in *aegyptiaca*. A long, curved, darkly pigmented process is present between the epandrium and gonocoxites, sometimes appearing to be connected with the epandrium and sometimes with the gonocoxites. It is definitely connected with the epandrium in species no. 1. Gonocoxites fused, with apical processes of specific form lateral to the dististyli in some species. The apical part with the dististyli and processes can be folded back, as in the genus *Geron*. The gonocoxites of some species bear a row of long or short spines at the apical margin. Dististyli triangular, more or less wide, with specific differences. Aedeagus long, tubular, with a wide opening situated at a right angle to its long axis. Aedeagal process long, V-shaped, of specific form. Apodeme triangular.

Spermathecae forming a strongly sclerotized spiral with 5–6 turns, which is bulb-shaped before entering the ducts. The ducts are wide to the ejection apparatus which is short,

81

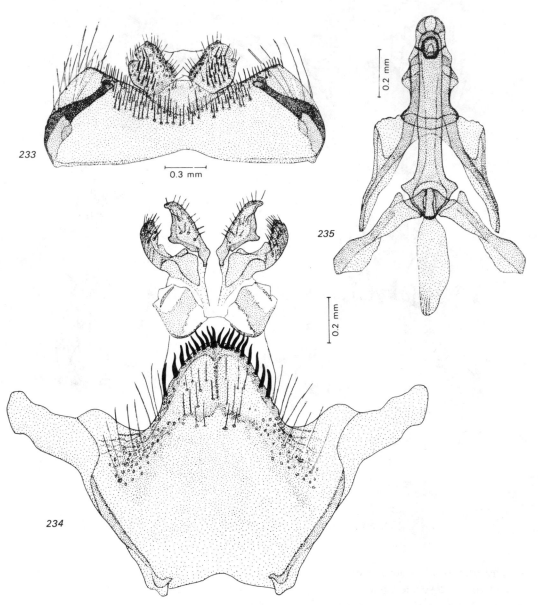

Figs. 233–235: *Toxophora* sp. no. 3 (Ethiopian)
233. epandrium; 234. gonopods; 235. aedeagus

striated, with small end plates. Furca U-shaped or with two separate bars having a sclerite of specific form between their apical ends. There is a large, rectangular sclerite behind the furca in some species. Eggs with an operculum.

T. epargyra was considered as a synonym of *maculata* by Efflatoun. This is incorrect. It is a distinct species, differing from *maculata* in coloration, wing venation and, particularly, in the the male genitalia (Figs. 215–218).

Lepidophora lepidocera (Wiedemann, 1828)

(Figs. 236–238)

Epandrium short, rounded posteriorly, with two spine-shaped processes which probably belong to the proctiger. Gonocoxites divided, broadly triangular. Dististyli parallel-sided, with a rounded and a pointed apical process. Aedeagus short, conical. Aedeagal process broadly rounded in lateral view. Hypandrium large, rounded.

Spermathecae very short, with large, flattened capsules with an indented apical part. Ducts short; ejection apparatus short, lacking processes but with large end plates. Furca with two bars which are widened posteriorly. Acanthophorites broad, with 70–100 long, curved spines with widened, curved apex in several rows.

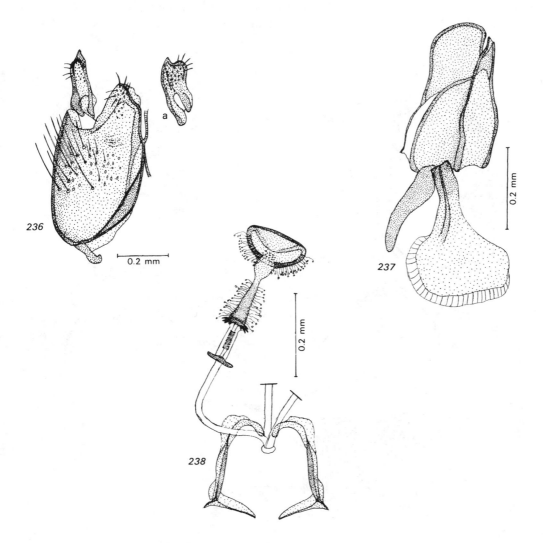

Figs. 236–238: *Lepidophora lepidocera*
236. gonopods; (a) dististylus, lateral; 237. aedeagus, lateral; 238. spermatheca

Cyrtomyia chilensis Paramonov, 1931
(Figs. 239 – 241)

The male genitalia resemble those of *Lepidophora*. Epandrium short, rounded posteriorly, with two pointed processes in the membrane at its base, which probably belong to the proctiger. Cerci triangular, Gonocoxites divided, broadly triangular, with a dense group of setae on the apical processes. Dististyli parallel-sided, with a pointed and a rounded apical process. Hypandrium large, triangular. Aedeagus short, conical, with a wide, rounded aedeagal process in lateral view.

Spermathecae short; capsules rhomboidal; ducts short, sclerotized, widening proximally. Ejection apparatus short, striated, with distinct end plates bearing processes. Furca with two bars, each with a lateral process at the basal end. Acanthophorites broad, partly connected with tergite 9, each with 20 – 30 long spines having curved ends in 4 – 5 rows.

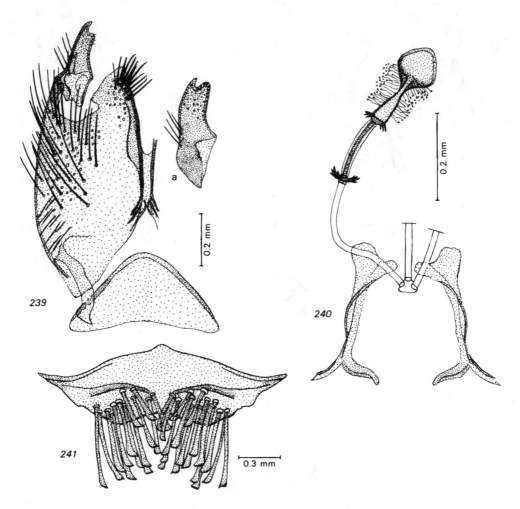

Figs. 239–241: *Cyrtomyia chilensis*
239. gonopods; (a) dististylus; 240. spermatheca; 241. tergite 9 of female with spines

Marmasoma sumptuosum White, 1907
(Fig. 242)

Only a female examined. Hull (1973) illustrated the male genitalia.

Epandrium rounded posteriorly, with long, proximal, lateral processes. Gonocoxites very elongate; dististyli parallel-sided, with a thin apical process. Aedeagus very long, thin, S-curved. Apodeme small, triangular [according to Hull (1973, Figs. 878–880)].

Spermathecae with small, ovoid capsules which continue gradually into a wide, sclerotized duct which narrows proximally. Ejection apparatus with small processes, lacking end plates. Ducts from the ejection apparatus to the vagina wide, striated in their proximal part. Furca V-shaped, with two wide, triangular bars. Tergite 8 invaginated, with a long, anterior apodeme. Tergite 9 with 9–12 ordinary spines. Acanthophorites absent.

The species resembles *Toxophora* in the development of long spines on the thorax but differs so distinctly in the male genitalia and the invaginated tergite 8 of the female that its position in this subfamily is doubtful. The spermathecae also differ distinctly from those of *Lepidophora* and *Cyrtomyia*, and the characteristic arrangement of the spines on tergite 9 of these species is absent.

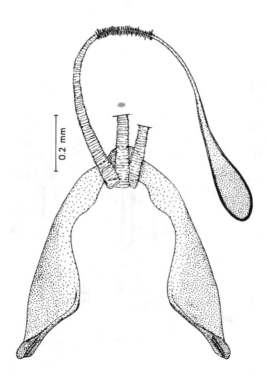

Fig. 242: *Marmasoma sumptuosum*, spermatheca

PHTHIRIINAE Becker, 1912

Phthiria Meigen, 1803

P. gaedei, vagans, xanthaspis, and several undetermined species (Palaearctic); *P. hilaris* (Australian) (Figs. 243–250)

Epandrium either short, rounded posteriorly or nearly triangular, with a small apical indentation (*vagans* group). Gonocoxites broad, with a short, inner apical process of

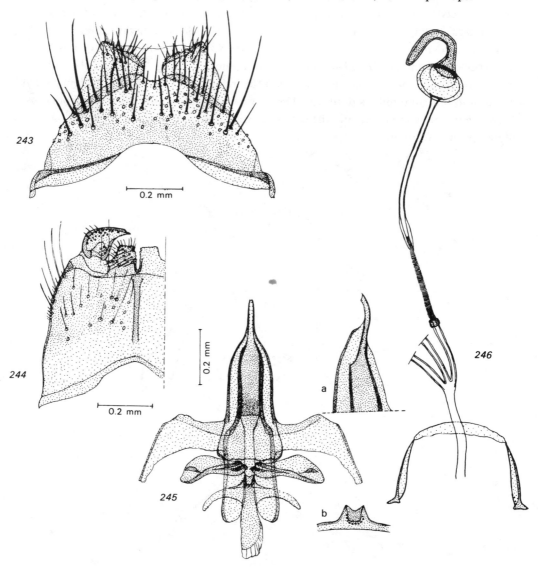

Figs. 243–246: *Phthiria gaedei*
243. epandrium; 244. gonopod; 245. aedeagus;
(a) same, apex, lateral; (b) head of apodeme; 246. spermatheca

specific form. Dististyli short, triangular, slightly curved. Aedeagus conical, without differentiations and lacking an aedeagal process. Basal plates small; apodeme small, with a process near its head.

Spermathecae with a rounded, slightly flattened, membranous capsule and a sclerotized, more or less long, recurved process. Ducts narrow, membranous. Ejection apparatus narrow, long, striated, without processes and with a small basal sclerotization. Ducts of *hilaris* shorter, sclerotized to near the ejection apparatus which has a large apical and a small basal end plate. Furca with two thin bars. Posterior sternite of the Old World species rectangular, with rounded posterior corners and two short setae close together in the middle of the posterior margin, oblong-trapezoidal, with a partly separated apical process bearing setae in *hilaris*.

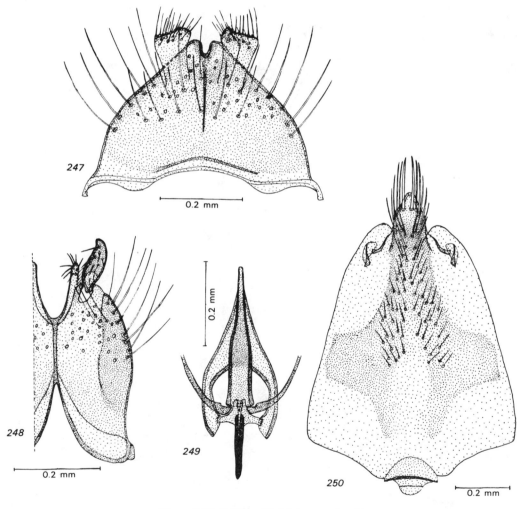

Figs. 247–249: *Phthiria xanthaspis*
247. epandrium; 248. gonopod; 249. aedeagus
Fig. 250: *Phthiria hilaris*, sclerite behind furca

Acreotrichus antecedens Walker, 1849

(Fig. 251)

Only a female examined, but Bowden (1971, p. 301) gives illustrations of the male genitalia.

Gonocoxites broad, with long apical processes. Dististyli long, narrow, wider apically. Aedeagus conical; sheath wide, with two short, curved, pointed apical processes.

Spermathecae as in *Phthiria*. Apical process of capsule very long. Basal part of capsule with opening of ducts widened, sclerotized. Ducts as in *Phthiria*. Ejection apparatus striated, with an apical and a basal end plate.

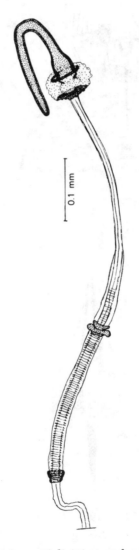

Fig. 251: *Acreotrichus antecedens*, spermatheca

ECLIMINAE Hall, 1969

The genera *Eclimus* and *Thevenemyia* were placed by Hull (1973) in the subfamily Bombyliinae. This is incorrect. They differ so distinctly from the other genera of the Bombyliinae that the establishment of the subfamily seems justified.

Eclimus gracilis Loew, 1844
(Figs. 252–256)

Epandrium short, trapezoidal, with two lateral, basal processes and two groups of 3–5 strong setae at the posterior corners. Gonocoxites broad, triangular, with two laterally

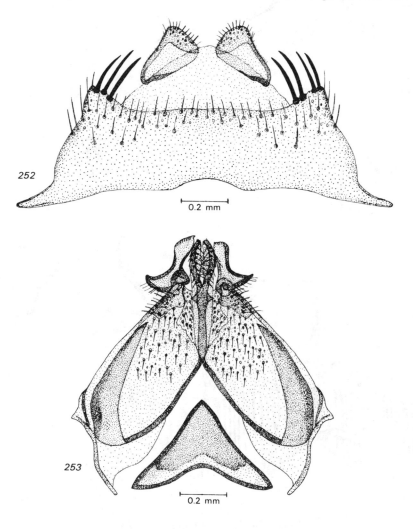

Figs. 252–253: *Eclimus gracilis*
252. epandrium; 253. gonopods

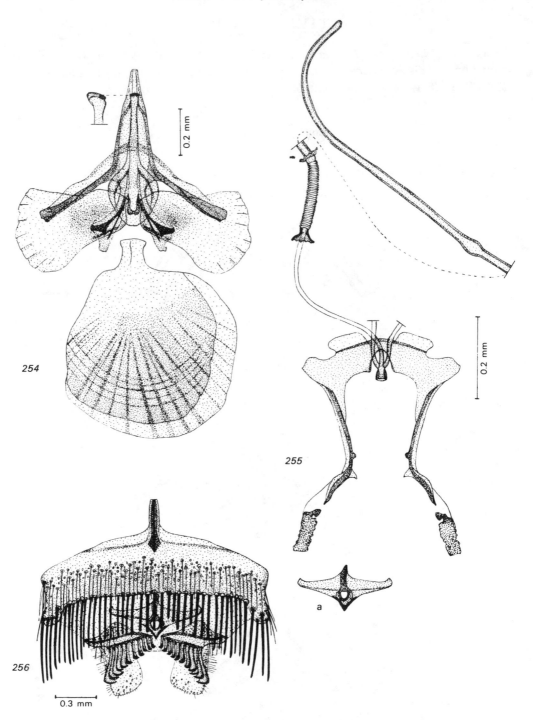

Figs. 254–256: *Eclimus gracilis*
254. aedeagus; 255. spermatheca;
256. tergite 8 and posterior tergites of female; (a) tergite 9

compressed apical processes. Dististyli rectangularly bent, with a pointed, lateral, basal process. Hypandrium large, triangular. Aedeagus conical; aedeagal process not reaching apex of aedeagus, with a club-shaped widening at the apex. Basal plates very large; apodeme very large, rounded.

Spermathecae forming long, narrow, sclerotized tubes which widen slightly and then narrow again before the ejection apparatus which is short, striated, without processes and with small end plates. Ducts to vagina short, sclerotized at the end, forming a short common duct. Furca rectangularly U-shaped, wider apically and basally. Tergite 8 only partly invaginated, with a long anterior apodeme and a dense row of long, strong setae posteriorly. Tergite 9 small, rhomboidal. Acanthophorites with 9–12 long spines with widened, hook-shaped ends.

Thevenemyia celer (Cole, 1919)
(Fig. 257)

Epandrium short, with rounded posterior corners and numerous thin setae in the posterior half. Gonocoxites as in *Eclimus* but apical inner processes shorter and wider. Hypandrium triangular, small. Dististyli triangular, but nearly parallel-sided in their greater basal part in other species. Aedeagus conical, short. Aedeagal process V-shaped, with rounded apex, extending slightly beyond apex of aedeagus.

Spermathecae as in *Eclimus*, with long, tubular capsules (end broken). Furca broadly U-shaped, with broad, basal inner processes. Tergite 8 as in *Eclimus*, with dense groups of long setae at the posterior corners and a longer row in the middle. Tergite 9 as in *Eclimus* but larger. Acanthophorites with 8–10 large spines having very wide, slightly curved ends. Median spines much smaller than the lateral ones.

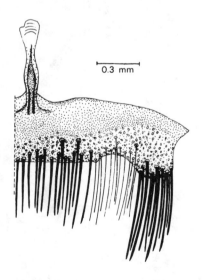

0.3 mm

Fig. 257: *Thevenemyia celer*, tergite 8 of female

91

SYSTROPINAE Brauer, 1880

Systropus Wiedemann, 1820

S. bicoloripennis, namaquensis, quadrinotatus (Ethiopian); *S. macer* (Nearctic) (Figs. 258 – 273)

The male genitalia are complicated and highly differentiated. They are not rotated, i.e., the epandrium is dorsal.

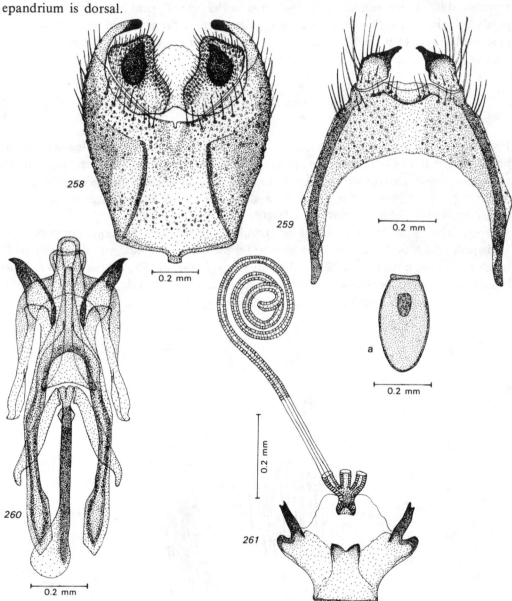

Figs. 258–261: *Systropus bicoloripennis*
258. epandrium; 259. gonopods; 260. aedeagus; 261. spermatheca; (a) egg

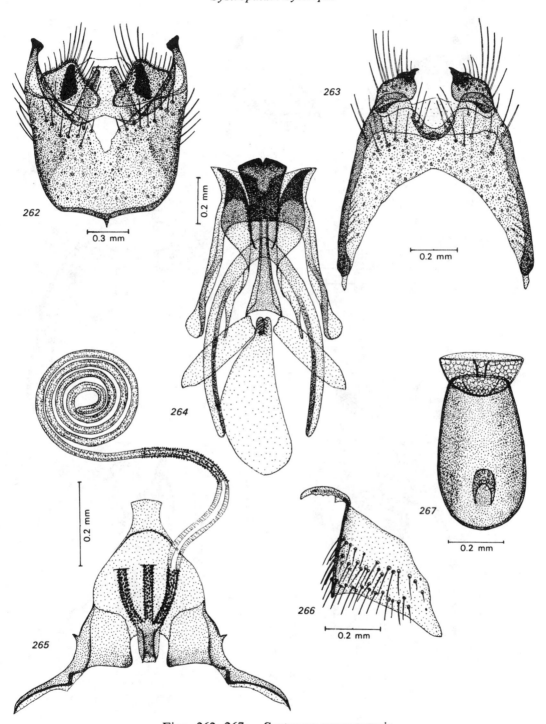

Figs. 262–267: *Systropus namaquensis*
262. epandrium; 263. gonopods; 264. aedeagus;
265. spermatheca; 266. tergite 8 of female, lateral; 267. egg

93

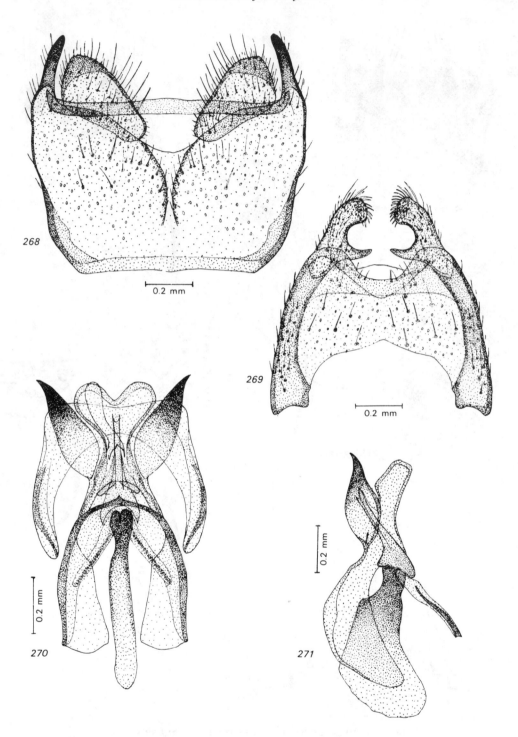

Figs. 268–271: *Systropus macer*
268. epandrium; 269. gonopods; 270. aedeagus; 271. same, lateral

Epandrium of specific form in every species examined, concave posteriorly, with long, lateral, posterior processes, partly divided in the American species *macer*. Cerci with a raised area of dense black tubercles in the Ethiopian species, which is absent in *macer*. Gonocoxites fused, with deeply concave basal margin. Dististyli triangular or nearly square, with a pointed apex which is black apically in the Ethiopian species, with a broadly rounded apex and a pointed basal process in *macer*. Sheath of aedeagus highly differentiated, and there are some structures the relationship of which is not clear. The aedeagal process is more or less wide, of specific form, extending beyond the apex of the aedeagus, very wide and indented apically in *macer*. There are two pointed lateral processes of the sheath which are usually black apically and extend beyond the aedeagus in some species. There is also a large, more or less U-shaped sclerite which is apparently not connected with the sheath. Its relationship will have to be studied in fresh material.

The spermathecae form dense, sclerotized spirals with 3–5 turns in most Ethiopian species, except in *macilentus* according to Hesse (1938, Fig. 309). Ducts wide, at first sclerotized, becoming striated and partly granulated, then wider and membranous. The part before the opening into the vagina is again granulated and forms a short common duct. A differentiated ejection apparatus is apparently absent. Furca triangular with a posterior concavity and processes on the basal ends, or of irregular form with processes in *bicoloripennis*. Tergite 8 partly invaginated, with a dorsally directed anterior apodeme. Eggs with a cup-shaped and reticulate operculum. There is also an elliptical sclerotized area in the middle or in the posterior half of the egg, apparently opening inwards, the function of which is not clear.

Hesse (1938, Fig. 320, p. 1016) gave an illustration of the spermathecae which he named 'coiled processes'. Mühlenberg (1971) gave a good photograph of the spermathecae of *S. celebensis* (Fig. 6, p. 11).

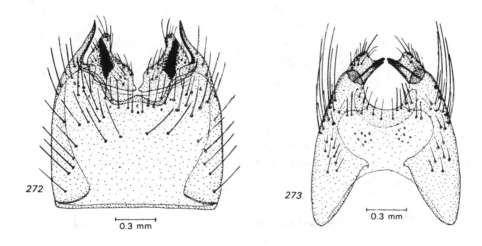

Figs. 272–273: *Systropus quadrinotatus*
272. epandrium; 273. gonopods

Dolichomyia Wiedemann, 1830

D. chilensis, coniocera (Figs. 274, 275)

Epandrium short, rounded posteriorly. Gonocoxites truncate-triangular, with rounded apex. Aedeagus conical, narrow. Aedeagal process with triangular, sclerotized, articulated part at the apex.

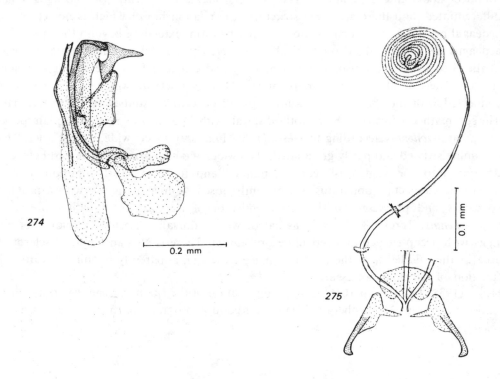

Fig. 274: *Dolichomyia coniocera*, aedeagus
Fig. 275: *Dolichomyia chilensis*, spermatheca

Spermathecae forming a small spiral with 4–5 turns. Ducts narrow, long. Ejection apparatus short, with flattened, cup-shaped end plates, lacking processes. Furca with two separate bars which are wider apically and with two posterior processes. A broad sclerite between the apical ends.

Zaclava flavifemorata Hull, 1973
(Fig. 276)

This Australian genus was placed in the past in the genus *Dolichomyia*.
Only a female examined. Spermathecae with long tubes which do not form a spiral. Ducts wide, covered by the gland, sclerotized, with a membranous, striated and granulated part

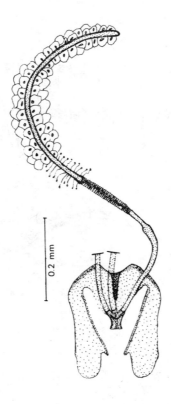

Fig. 276: *Zaclava flavifemorata*, spermatheca

in the middle. End of ducts and short common duct sclerotized. Furca U-shaped, with a large, triangular, inner process at the apex.

BOMBYLIINAE Latreille, 1862

This is a very large subfamily, but Hull (1973) included in it a number of genera which obviously do not belong to it. The Ecliminae were removed by Hall (1969). The *Corsomyza* group and the *Cytherea* group belong to the Tomophthalmae according to the structure of the head, and the *Crocidium* group probably also does not belong to the Bombyliinae.

Bombylius Linné, 1758

About 30 Palaearctic species and two Nearctic species of the subgenus *Zephyrectes* (*anthophoroides, montanus*) (Figs. 1, 8, 277–290)

Male genitalia simple, without distinct differentiations, but with specific differences in form. Epandrium either rectangular, with rounded posterior corners, or with rounded posterior margin and obliquely posteriorly directed corners. Gonocoxites truncate-

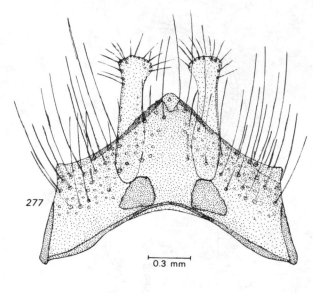

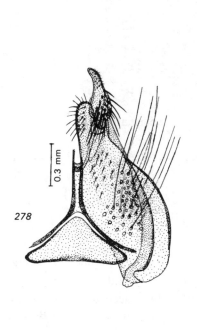

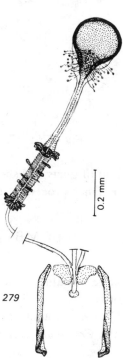

Figs. 277–279: *Bombylius major*
277. epandrium; 278. gonopod; 279. spermatheca

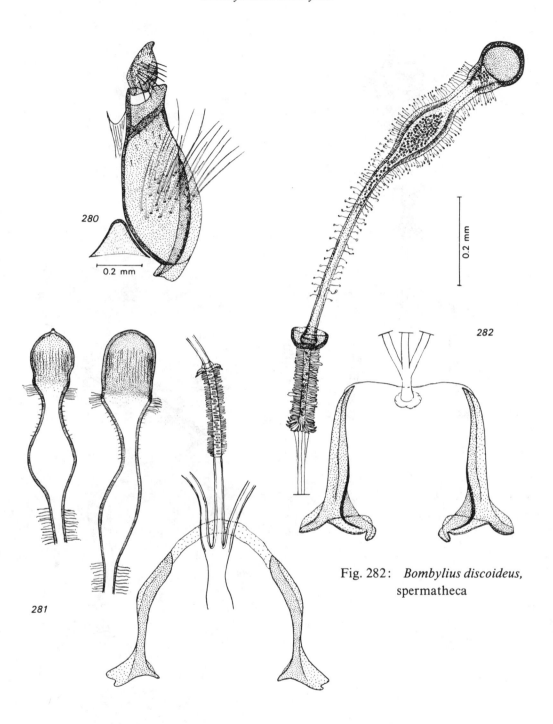

280

0.2 mm

282

0.2 mm

281

Fig. 282: *Bombylius discoideus,* spermatheca

Figs. 280–281: *Bombylius ater*
280. gonopod; 281. spermathecae

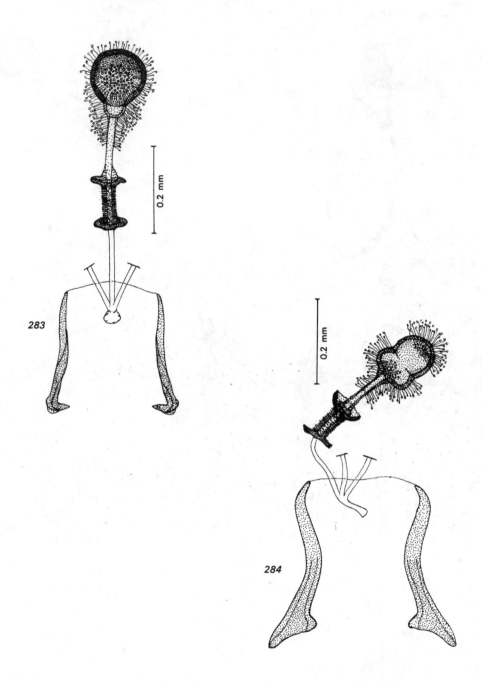

Fig. 283: *Bombylius argentifrons*, spermatheca
Fig. 284: *Bombylius androgynus*, spermatheca

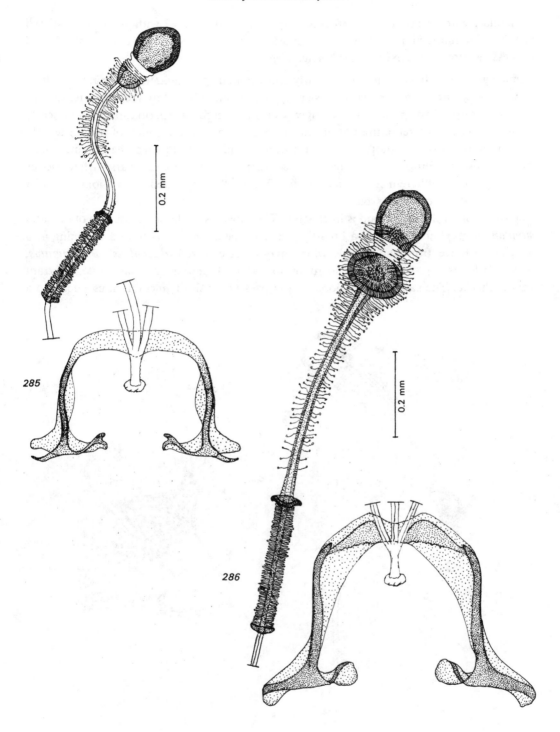

Fig. 285: *Bombylius fuscus*, spermatheca
Fig. 286: *Bombylius punctatus*, spermatheca

triangular, with an apical inner process of specific form, with or without setae. Dististyli usually triangular, more or less narrow and long. Aedeagus simple, without an aedeagal process, except in the Nearctic subgenus *Zephyrectes*.

Female genitalia. In contrast to the nearly uniform male genitalia, the spermathecae show great specific variation, even in species groups well defined by external characters, such as the *ater* group. Most species have strongly sclerotized, globular sperm capsules, but nearly all show specific differentiations at or near the base of the capsule. Ducts of varying length; ejection apparatus with short and larger processes and two end plates (the proximal plate may be absent). Furca either U-shaped or consisting of two separate bars. A few species have capsules of different form (*androgynus*). Tergite 8 invaginated, with a long, anterior apodeme with a triangular base.

The situation in the *ater* group is interesting. The capsules of the spermathecae of *ater* and *simulans* are cylindrical, with a broadly rounded apex. The ducts have a widening at a varying distance from the capsule which may be sclerotized (*discoideus*). *B. punctatus*, which belongs to this group according to external characters, has quite aberrant spermathecae. The capsules are globular and pass via a short, membranous duct into a

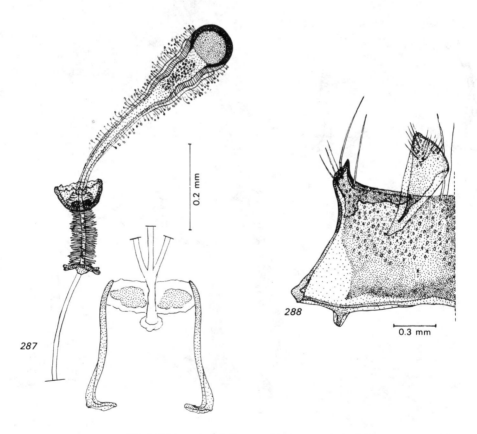

Fig. 287: *Bombylius vulpinus*, spermatheca
Fig. 288: *Bombylius (Zephyrectes) anthophoroides*, epandrium

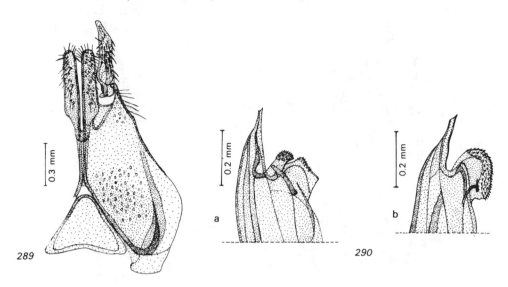

Fig. 289: *Bombylius (Zephyrectes) montanus*, gonopod
Fig. 290: *Bombylius (Zephyrectes)*, aedeagal process ;
(a) *B. montanus*; (b) *B. anthophoroides*

second globular, slightly flattened, also strongly sclerotized capsule which is surrounded by the gland. The spermathecae of *fuscus* have a similar development but the additional sclerotization is much smaller and does not form a complete capsule.

The differentiation of the spermathecae thus does not agree with the external characters and is apparently a specific development.

The species of the subgenus *Zephyrectes* have a large aedeagal process with small denticles on the sheath, which has not been found in any of the Old World species examined (Figs. 288 – 290). Spermathecae as in *Bombylius*, with specific differences.

Parabombylius subflavus Painter, 1926

The male genitalia closely resemble those of some species of *Bombylius*, with differences which do not seem to be of generic character.

The spermathecae are also similar to those in some species of *Bombylius* (e.g., sp. near *venosus*, Fig. 8), with globular capsules and long ducts.

Anastoechus Osten-Sacken, 1877

A. bahirae, exalbidus, sp. near *hyrcanus, nitidulus, niveus, stramineus, trisignatus* (Figs. 291 – 298)

Epandrium short, curved, with rounded posterior margin and concave anterior margin or nearly rectangular, exhibiting specific differences in form. Gonocoxites truncate-triangular, sometimes narrow, nearly parallel-sided, with apical inner processes of specific form.

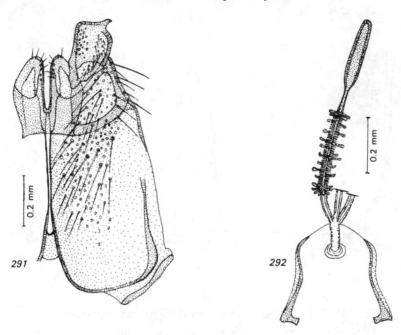

Figs. 291–292: *Anastoechus exalbidus*
291. gonopod; 292. spermatheca

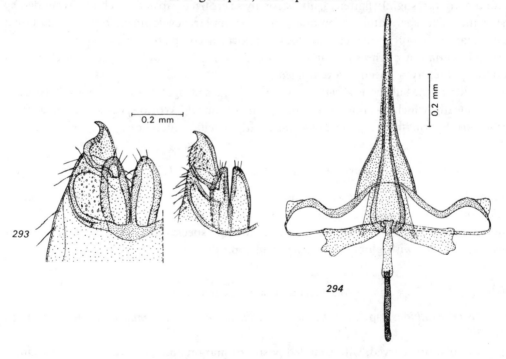

Figs. 293–294: *Anastoechus stramineus*
293. gonopods, apex; 294. aedeagus

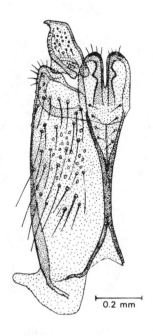

Fig. 295: *Anastoechus nitidulus*, gonopod

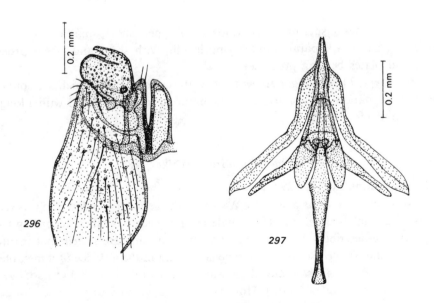

Figs. 296–297: *Anastoechus niveus*
296. gonopod; 297. aedeagus

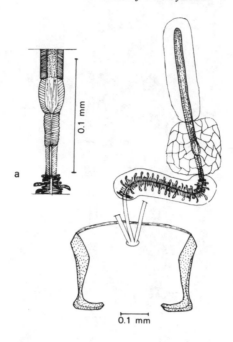

Fig. 298: *Anastoechus bahirae*, spermatheca; (a) basal part of duct

Dististyli either triangular, with rounded sides, or broad, nearly square, with a pointed apical process (*exalbidus, niveus*). Hypandrium small, triangular. Aedeagus without differentiations, but of varying form, either long and conical or short. Sheath with lateral bulges.

There are two types of spermathecae. Capsules of type 1 short, spindle-shaped; ducts very short, thin. Ejection apparatus of varying length, with short and larger processes and distinct end plates bearing processes.

Capsules of type 2 form long, narrow tubes with rounded apex, which are connected with the ejection apparatus by a short, wide, membranous duct. Tergite 8 with a long, narrow, anterior apodeme.

Choristus bifrons Walker, 1852

One female examined (Fig. 299)

This Australian species was placed by Walker in the genus *Bombylius* and was removed by Roberts (1928) to *Anastoechus*. The female has not been described according to Bowden (1971), who re-described the male, but not the genitalia. He stated that it differs from *Anastoechus* and *Systoechus* in the wing venation and in the absence of a metapleural tuft. The female resembles *Anastoechus* in the wide frons (width of head 3 mm, of eye 0.65 mm, of frons near antennae 1.65 mm). However, the frons and face are much longer than in *Anastoechus*, sparsely pilose, distinctly projecting and widening ventrally.

The spermathecae differ distinctly from those of *Anastoechus*. The sperm capsules are small, ovoid; the ducts are very long, thick and sclerotized in the apical half and then

106

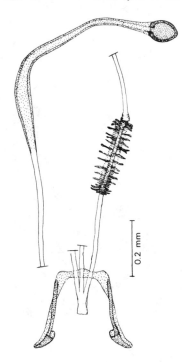

Fig. 299: *Choristus bifrons,* spermatheca

become very narrow. Ejection apparatus short, strongly sclerotized, with short and longer processes and without end plates but with longer and curved processes at both ends. The furca has two separate bars which are widened at the base. Acanthophorites with 8–10 thick, long, curved spines.

Systoechus Loew, 1855

S. gradatus, pallidipilosus, sulphureus; S. albiceps (Australia) probably belongs to another genus (Figs. 300–310)

Epandrium short, nearly rectangular. Gonocoxites broad, truncate-triangular. Dististyli narrow, triangular. Hypandrium small, triangular. Aedeagus conical, with inner differentiations in the base of the apical part. Sheath of aedeagus with complicated differentiations. It has a large basal bulge from which extend two long, apical processes with a slightly widened, pointed apical part and two inner, finger-shaped processes between which is situated a narrow, pointed or bifid process resembling the aedeagus in outline. The aedeagus of some Ethiopian species illustrated by Hesse (1938) is distinctly different. It is S-curved and has a large, laterally compressed keel.

The spermathecae of the Palaearctic species examined have a small, rounded capsule with or without a small apical process. Ducts at first wide, sclerotized, then narrowing to the ejection apparatus which is moderately long, with short and longer processes and small end plates. Furca with two separate bars which are widened posteriorly and have an inner

107

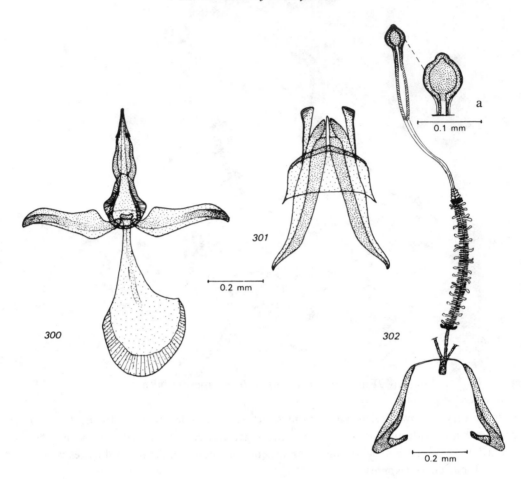

Figs. 300–302: *Systoechus sulphureus*
300. aedeagus ; 301. aedeagal process ; 302. spermatheca, (a) capsule enlarged

process at the base. Tergite 8 with a long anterior apodeme with a broad, triangular base. Acanthophorites with 8–9 long spines.

The spermathecae of the Australian species determined as *S. albiceps* by Paramonov differ distinctly from those of the Palaearctic species examined (Fig. 310). The sclerotized, apical part of the sperm capsule is conical, while the basal part is membranous and broadly rounded. Ducts short, at first narrow, then widening to the ejection apparatus which has no processes but only small tubercles and no end plates. Ducts to the vagina long, with a short common duct. Furca with two separate bars which widen posteriorly and bear two pointed inner processes in the posterior half. There is a broadly V-shaped sclerite between the apical ends of the bars.

The species resembles *Systoechus* in wing venation but differs from it in the form of the head, the very narrow frons and other characters, and probably belongs to another genus. Male not examined.

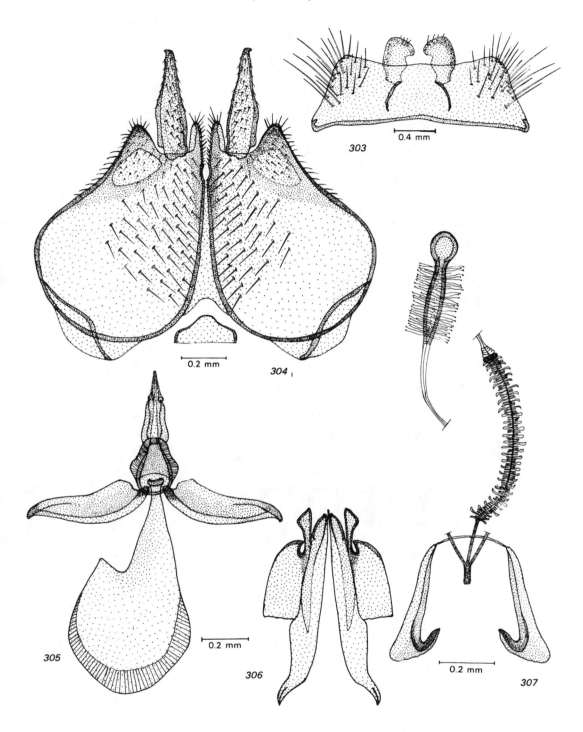

Figs. 303–307: *Systoechus gradatus*
303. epandrium; 304. gonopods; 305. aedeagus; 306. aedeagal process; 307. spermatheca

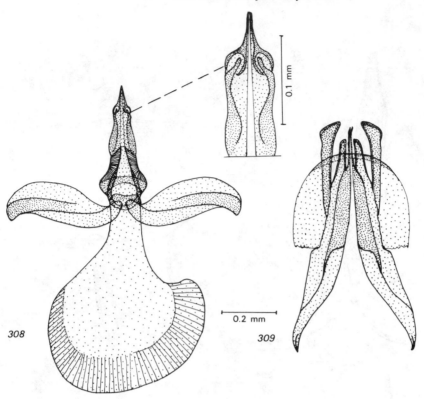

Figs. 308–309: *Systoechus pallidipilosus*
308. aedeagus and its apex, enlarged; 309. aedeagal process

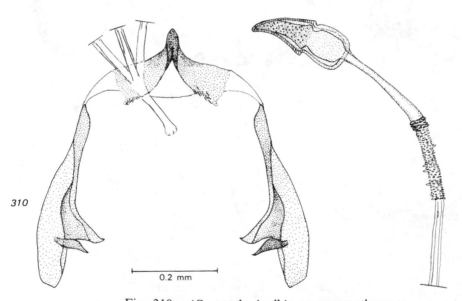

Fig. 310: *'Systoechus' albiceps*, spermatheca

The above characters and the distinctly different male genitalia illustrated by Hesse suggest that several genera have probably been placed in *Systoechus* and a revision of this genus seems necessary.

Dischistus Loew, 1856

D. hirticeps, mystax, plumipalpis (Ethiopian); *D. (Acanthogeron) senex, separatus, syriacus* (*Lissomerus* Austen is a synonym of *Acanthogeron*); '*Dischistus*' *melampogon* (Chile) (Figs. 311 – 323)

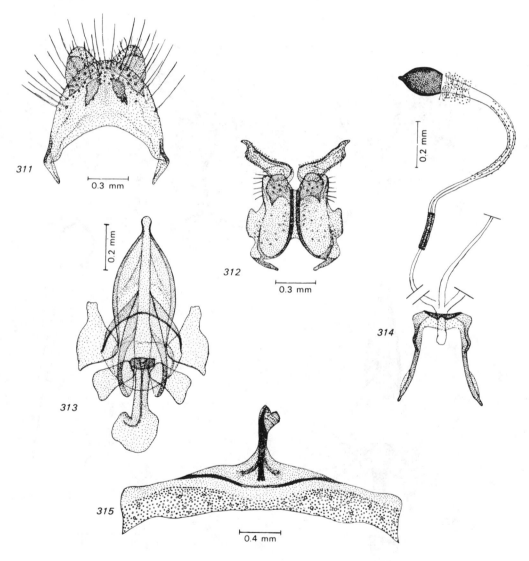

Figs. 311–315: *Dischistus mystax*
311. epandrium; 312. gonopods; 313. aedeagus;
314. spermatheca; 315. tergite 8 of female

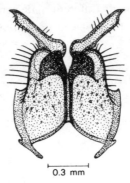

Fig. 316: *Dischistus hirticeps*, gonopods

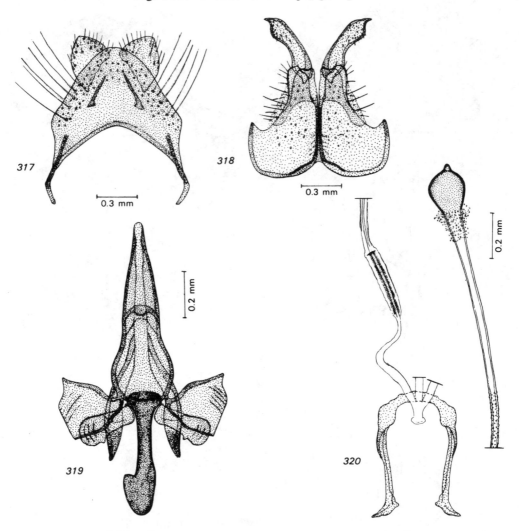

Figs. 317–320: *Dischistus (Acanthogeron) syriacus*
317. epandrium; 318. gonopods; 319. aedeagus; 320. spermatheca

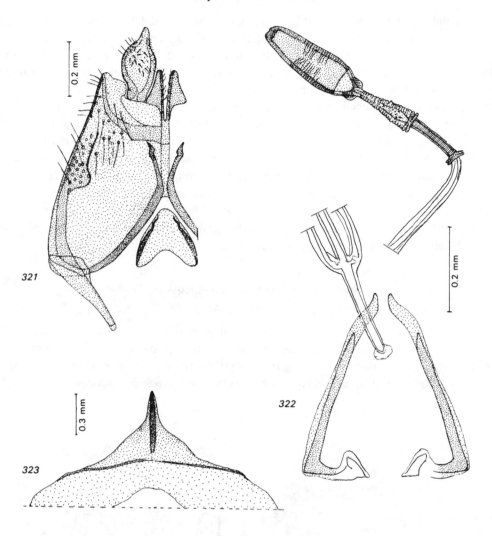

Figs. 321–323: *'Dischistus' melampogon* (Chile)
321. gonopods; 322. spermatheca; 323. tergite 8 of female, apical part

The genus was based on the South African species *mystax*. The Palaearctic species later placed in this genus belong to the genus *Bombylisoma* Rondani with the type species *minimus* Schrank, 1781. The genus *Acanthogeron* differs from *Dischistus* only in the closed and stalked cell r_5 but resembles it in all other characters and thus may be considered as a subgenus of *Dischistus*.

Epandrium broadly rounded posteriorly, deeply concave anteriorly. Gonocoxites broadly truncate or wide posteriorly and narrow in the apical half and with broadly rounded, inner apical processes. Dististyli narrow, long, nearly parallel-sided, with a small, pointed apical process, similar in all species, with only specific differences. Aedeagus conical, or with convex sides. Aedeagal process absent; apodeme narrow, curved.

Spermathecae with globular or ovoid capsules bearing a small apical process. Ducts long, at first wide, then narrowing. Ejection apparatus narrow, with small processes and lacking end plates, with specific differences in the length of the ducts and the ejection apparatus. Furca U-shaped, with basal lateral processes in some species. Tergite 8 with a long anterior apodeme which is laterally compressed and bears a vertical ridge.

The Chilean species placed by Hall (1975) in this genus certainly do not belong to it. In the past they were placed in the genus *Sparnopolius*, but they may not belong to this genus either. The spermathecae are quite different (Fig. 322). The male genitalia resemble those of *Bombylius*. The spermathecae of *D. melampogon* have oblong-conical capsules with a separate basal sclerotization. Ducts very short, widening proximally. Ejection apparatus very short, striated, without processes and with small end plates. Ducts to vagina very long; common duct moderately long. Furca V-shaped, with inner basal processes. Tergite 8 with a large triangular apodeme, closely resembling that of *Sparnopolius* (Figs. 321–323). Acanthophorites with about 12 thick, curved spines.

Sparnopolius lherminierii (Macquart, 1840)
(Fig. 324)

Epandrium short, nearly rectangular, rounded posteriorly, concave anteriorly, with a distinct, small indentation in the middle of the posterior margin. Gonocoxites truncate-triangular; dististyli oblong-triangular, with rounded apex. Hypandrium short, crescent-shaped. Aedeagus conical, with convex sides. Aedeagal process absent.

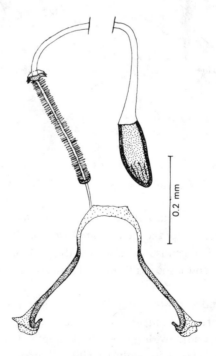

Fig. 324: *Sparnopolius lherminierii,* spermatheca

Spermathecae with oblong-ovoid, slightly conical capsules. Ducts very long, membranous. Ejection apparatus short, with short processes and small end plates. Furca V-shaped; bars widening posteriorly. Tergite 8 with a triangular apodeme. Acanthophorites with about 20 long, thin setae with curved apex.

New Genus
(Figs. 325–327)

The species *efflatounbeyi* Francois, 1961 (= *auripilus* Efflatoun, 1945, not of Séguy) and *blanchei* Efflatoun, 1945 are usually placed in the genus *Acanthogeron*, but they certainly

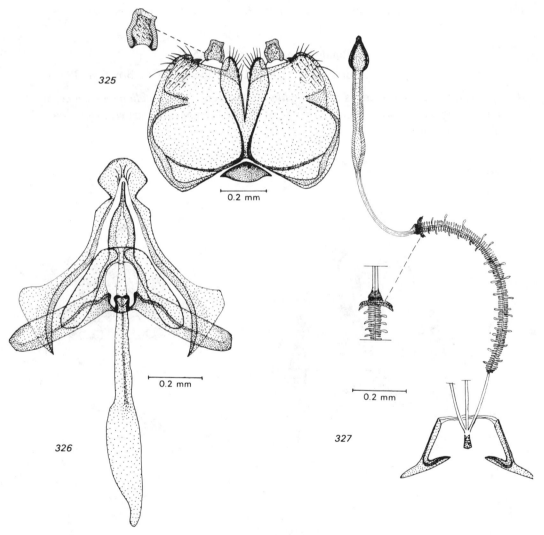

Figs. 325–327: (?) *'Acanthogeron' efflatounbeyi* Francois (*auripilus* Efflatoun)
325. gonopods; 326. aedeagus; 327. spermatheca

115

do not belong to this genus. The antennae are different; segment 1 is short and narrow, not long and thickened as in the species of *Dischistus*. The genitalia are also quite different.

Epandrium rectangular, with rounded posterior corners and deeply concave anterior margin. Gonocoxites very broad, nearly rectangular. Dististyli very short, rectangular, with a small, lateral, apical point. Hypandrium small, broadly triangular. Aedeagus conical. Aedeagal process V-shaped, with widened apical part which extends distinctly beyond the apex of the aedeagus.

Spermathecae with small, pear-shaped, pointed capsules, Ducts at first wide, then narrowing. Ejection apparatus long, narrow, with short processes and relatively few larger processes and with a distinct apical end plate. Furca with two bars with an outer and an inner basal process. Acanthophorites with 5–6 long, thick, curved spines.

Bombylisoma Rondani, 1856

(*Dischistus* auct.) (= *Chasmoneura* Hesse, 1938, according to Bowden, 1973)

B. brevirostratum (described as a *Bombylius* by Austen), *breviusculum, melanocephalum, minimum trigonum* (= *D. pulchellus* Austen); *Chasmoneura argyropyga, coracina, cinereicinta, pectoralis* (Figs. 328–333)

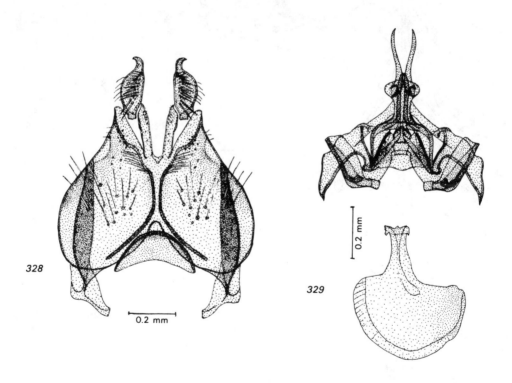

Figs. 328–329: *Bombylisoma minimum*
328. gonopods; 329. aedeagus

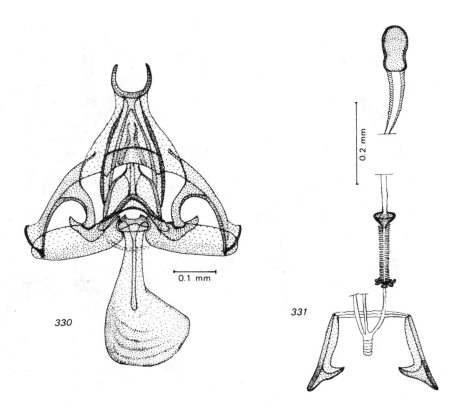

330

331

Figs. 330–331: *Bombylisoma trigonum*
330. aedeagus; 331. spermatheca

Epandrium short, rectangular, with more or less concave anterior and posterior margin, sometimes with strong setae at the posterior margin (*cinereicincta*). Gonocoxites truncate-triangular. Dististyli triangular, more or less wide, their basal part sometimes with dense setae (*pectoralis*) or a ridge with dense setae.

Hypandrium large, triangular, with rounded apex. Aedeagus conical, with more or less wide base, or very short, with wide, rounded base (*coracina*). Sheath with complicated processes. Aedeagal process either long, extending beyond apex of aedeagus (*trigonum*) and with two narrow processes, or processes are situated on the basal part of the sheath. They vary widely in length and form, being either short and curved or long, thin, slightly curved and twisted.

Spermathecae with small capsules of varying form, ovoid or club-shaped (*trigonum*). Ducts long, thin, sometimes widened near the capsule. Ejection apparatus of varying length, with only small processes and small end plates. Furca with two separate bars which are widened posteriorly and may have an inner posterior process. Tergite 8 with a long apodeme with a triangular base. Acanthophorites with 8–12 long spines.

117

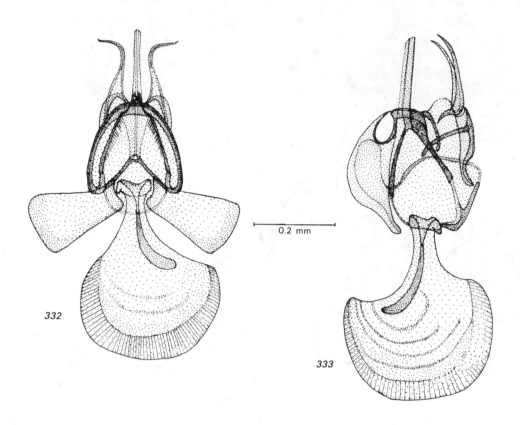

Figs. 332–333: *Bombylisoma breviusculum*
332. aedeagus, dorsal; 333. same, lateral

Doliogethes Hesse, 1938

D. luridus, pallidulus, pullatus, seriatus, tripunctatus, vagans (Figs. 334–337)

Epandrium trapezoidal or rectangular, with concave anterior margin. Gonocoxites truncate-triangular, or very wide, rounded basally, with small, inner apical processes. Dististyli triangular, more or less wide, with short hairs. Hypandrium triangular. Aedeagus conical, with wide base. Aedeagal process absent (*pullatus*) or forming two long, apical processes with slightly widened apex (*luridus, pallidulus*).

Spermathecae with small, globular or ovoid capsules, which are very small in *luridus*. Ducts more or less long, at first wide, sclerotized, then narrowing proximally. Ejection apparatus long, narrow, with short and longer processes, lacking end plates. Furca either U-shaped (*luridus*) or consisting of two bars, their basal ends widened into an inner and an outer process. Tergite 8 with a long apodeme having a triangular base. Acanthophorites with 4–6 long spines.

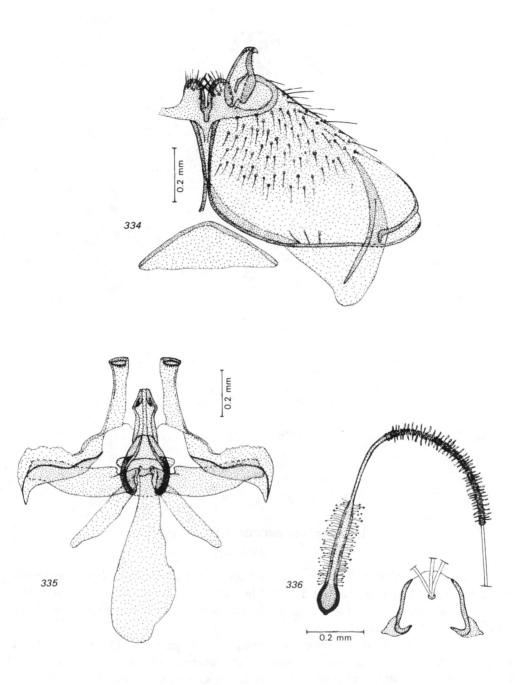

Figs. 334–336: *Doliogethes pallidulus*
334. gonopods; 335. aedeagus; 336. spermatheca

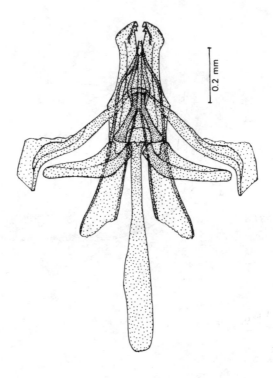

Fig. 337: *Doliogethes luridus*, aedeagus

Lepidochlanus fimbriatus Hesse, 1938
(Figs. 338–341)

Epandrium trapezoidal, with deeply indented posterior margin. Gonocoxites triangular, with narrow, inner apical processes. Hypandrium absent. Dististyli narrow, parallel-sided, with a short apical point. Aedeagus short, conical, simple; sheath with lateral shoulders. Aedeagal process absent.

Spermathecae with small, globular or slightly conical capsules. Ducts at first wide, sclerotized, then narrowing, twice as long in one specimen than in a second examined. Ejection apparatus short, with short and longer processes, lacking end plates, but with longer processes at the ends. Furca with two bars. Posterior segments of the abdomen not differentiated. Tergite 8 not invaginated. Tergite 9 without acanthophorites or spines. This is a quite exceptional feature in the Bombyliinae. A similar situation but with special differentiations has been noted only in the *Lordotus* group (see p.130).

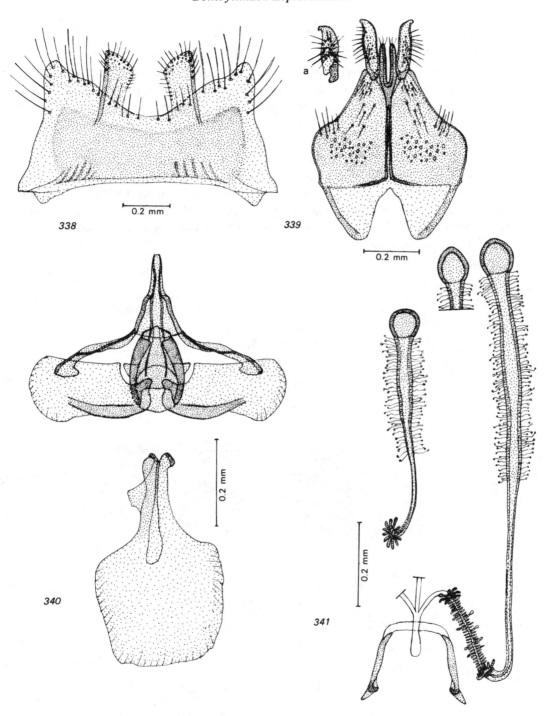

Figs. 338–341: *Lepidochlanus fimbriatus*
338. epandrium; 339. gonopods; (a) dististylus, lateral;
340. aedeagus; 341. spermatheca in two specimens

Gonarthrus Bezzi, 1924

G. namaënsis and an undetermined species from Kenya (Figs. 342–347)

Epandrium long, rounded posteriorly, its basal margin deeply concave and with long basal processes which are sometimes several times longer than the median part of the epandrium. Gonocoxites triangular. Dististyli of characteristic form, club-shaped, slightly curved, with a small pointed process near the apex and a dense group of strong setae on the dorsal side. Aedeagus long, narrow, conical, slightly S-curved. An aedeagal process is absent but the sheath has a short, club-shaped process near the middle of the aedeagus.

Spermathecae with small, ovoid capsules, with more or less pointed apex. Ducts moderately long, wider near the capsules. Ejection apparatus very long, narrow, with small processes, without end plates. Furca with two bars which are slightly wider posteriorly. Tergite 8 with a wide, triangular apodeme. Tergite 9 with 3–4 spines. Acanthophorites only indistinctly separated from tergite 9.

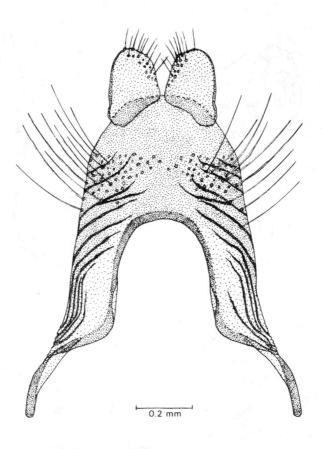

0.2 mm

Fig. 342: *Gonarthrus namaënsis*, epandrium

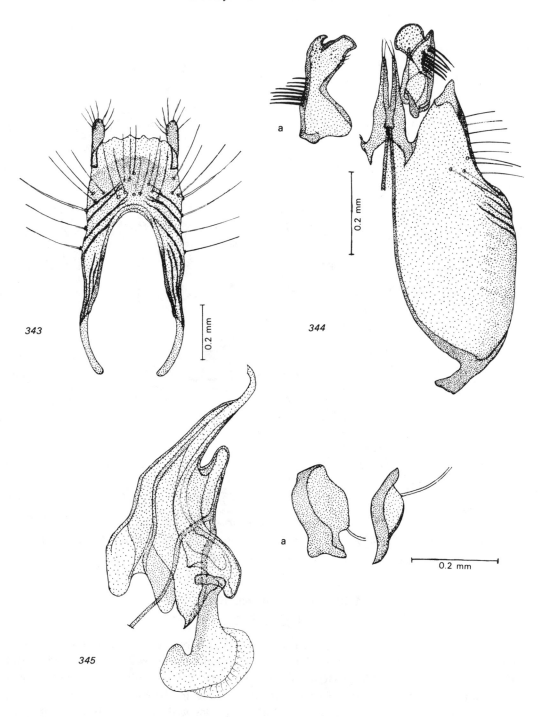

Figs. 343–345: *Gonarthrus* sp. (Kenya)
343. epandrium; 344. gonopods; (a) dististylus, lateral; 345. aedeagus;
(a) different aspects of lateral plates

123

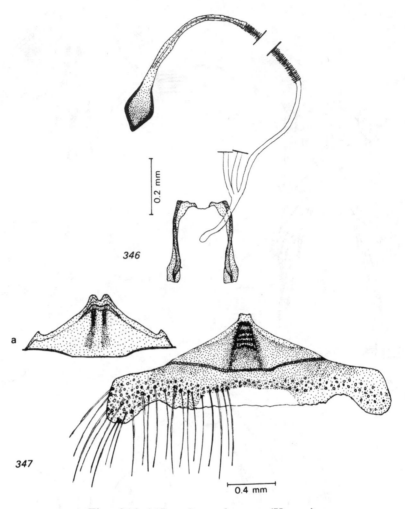

Figs. 346–347: *Gonarthrus* sp. (Kenya)
346. spermatheca; 347. tergite 8 of female of *G. namaënsis* from South Africa;
(a) apodeme of tergite 8 of species from Kenya.

Efflatounia aegyptiaca Bezzi, 1925
(Figs. 348–351)

Epandrium short, with projecting lateral posterior corners and sinuate posterior margin. Gonocoxites truncate-triangular. Dististyli triangular, narrow, with curved apex. Hypandrium absent. Aedeagus conical. Aedeagal process laterally compressed, with a long, curved, pointed ventral process with an opening in the middle.

Spermathecae very short, with large, globular capsules with a conical base. Ducts very short, sclerotized. Ejection apparatus also short, with short processes and without end plates. Furca with two bars which are widened posteriorly. Tergite 8 with a long apodeme having a triangular base. Acanthophorites with 10 long, thin setae.

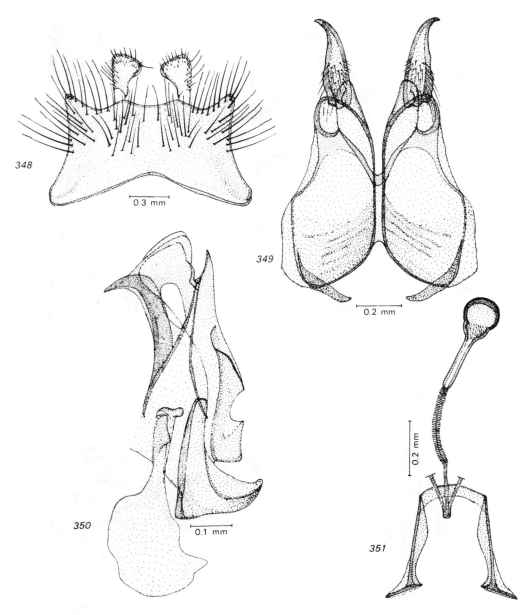

Figs. 348–351: *Efflatounia aegyptiaca*
348. epandrium; 349. gonopods; 350. aedeagus, lateral; 351. spermatheca

Eurycarenus Loew, 1860

E. dichopticus, laticeps, and an undetermined species from Kenya (Figs. 352–355)

Epandrium rectangular, with rounded or angular posterior corners. Gonocoxites truncate-triangular, with an apical process which bears one or two rows of short black

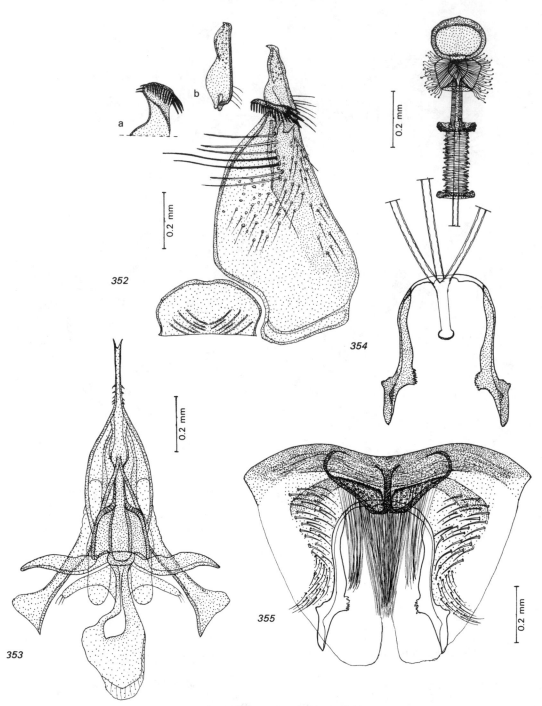

Figs. 352–355: *Eurycarenus laticeps*
352. gonopods; (a) apical process of gonopods, different aspect; (b) dististylus of
E. dichopticus; 353. aedeagus; 354. spermatheca of *Eurycarenus* sp.;
355. sclerite behind furca of *Eurycarenus* sp.

spines and longer spines laterally. Dististyli long, narrowly triangular or with slightly widened apex and a small, curved apical point. Hypandrium pentagonal or nearly semicircular. Aedeagus long, conical, with long, narrow apical part. Several denticles at the base of the narrow apical part. Aedeagal process absent.

Spermathecae with slightly flattened, globular capsules behind which is a membranous part. Ducts very short. Ejection apparatus also very short, with only small processes. End plates large, shallow, cup-shaped. Furca U-shaped, with widened posterior ends. Tergite 8 with a long apodeme with a triangular base. There is a large sclerite of characteristic form, truncate-triangular, with a triangular sclerotization near its anterior part and a thick brush of long setae below the furca. Acanthophorites with 15–20 long spines with curved apex.

Heterostylum robustum Osten-Sacken, 1877
(Figs. 356–358)

Epandrium rounded posteriorly, with deeply concave anterior margin. Gonocoxites triangular. Hypandrium small, rounded-triangular. Dististyli deeply bifid, with a club-shaped apex, a conical subapical process and a process with rounded apex at the base. Aedeagus long, conical, with wide base. Aedeagal process very long, with curved, pointed apical part.

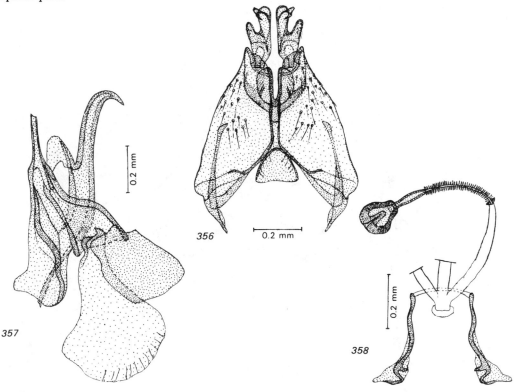

Figs. 356–358: *Heterostylum robustum*
356. gonopods; 357. aedeagus, lateral; 358. spermatheca

Spermathecae short; capsules nearly globular, with very thick walls and a large, inner, finger-shaped process at the apex. Ducts very short, sclerotized. Ejection apparatus longer, narrow, with only small processes and a small basal end plate. Ducts to vagina wide, membranous. Furca with two thin bars which are widened posteriorly. Tergite 8 with a short, narrow apodeme. Acanthophorites with about 10 long spines.

Triploechus novus (Williston, 1893)
(Figs. 359–361)

Epandrium short, rectangular, slightly curved. Gonocoxites with narrow apical part and wider basal part. Inner apical processes long, narrow, with hairs at the apex. Dististyli narrowly triangular, slightly curved. Hypandrium very small, rounded-triangular. Aedeagus broadly conical. Aedeagal process long, laterally compressed, with a long, pointed, slightly curved apical process and a long, narrow opening in the middle, resembling that of *Efflatounia*.

Spermathecae very short. Capsules globular, with conical base. Ducts very short, sclerotized. Ejection apparatus short, with small processes, lacking end plates, closely resembling those of *Efflatounia*. Furca with two bars, the posterior part of which is widened. Tergite 8 with a narrow apodeme with a triangular base. Acanthophorites with 4–6 long spines.

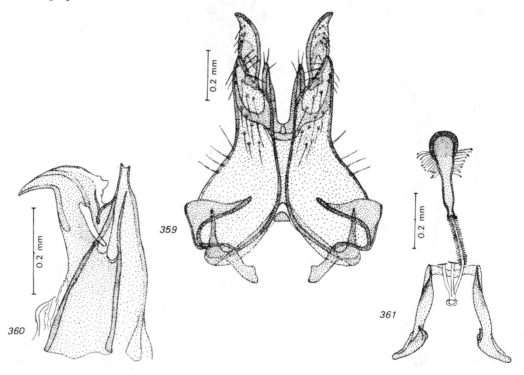

Figs. 359–361: *Triploechus novus*
359. gonopods; 360. aedeagus, apex, lateral; 361. spermatheca

Bombyliinae : Hallidia

Hallidia plumipilosa Hull, 1970
(Figs. 362–364)

Epandrium short, rectangular, with projecting posterior corners. Gonocoxites triangular. Hypandrium small, triangular. Dististyli oblong-oval, wider apically, with a short point at the apex. Aedeagus long, pointed, slightly S-curved. Sheath short, with a short apical process. Aedeagal process absent.

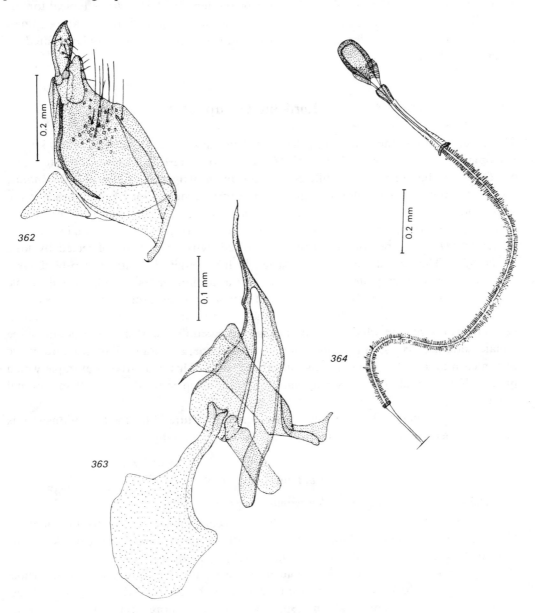

Figs. 362–364: *Hallidia plumipilosa*
362. gonopods; 363. aedeagus, lateral; 364. spermatheca

Spermathecae with small, club-shaped capsules and a slightly conical sclerotization at the beginning of the ducts which are short and membranous. Ejection apparatus very long, narrow, with short processes and small end plates. Furca not recognizable. Tergite 8 invaginated, with a long apodeme with a triangular base. Acanthophorites distinct, with 7–8 widely spaced, short spines.

Hall (1975) considered *Hallidia* as intermediate between *Lordotus* and *Geminaria*. However, the differentiation of the abdomen of the female of *Hallidia* is typical for the Bombyliinae, with an invaginated tergite 8 and distinct acanthophorites bearing spines, while the abdomen of the females of *Lordotus* and *Geminaria* is quite undifferentiated, as described below.

Lordotus Group

The group contains the American genera *Lordotus* and *Geminaria* and the little known Ethiopian genus *Othniomyia* Hesse, 1938. *Lordotus* closely resembles *Bombylius* externally, but *Geminaria* has a completely different habitus, resembling some species of *Aphoebantus*, also in its shining, black, bilobed scutellum. The three genera resemble each other in the characteristic wing venation with three submarginal cells. Examination of the female abdomen showed that it has a quite aberrant structure. Tergite 8 is not invaginated. The posterior segments are normal, slightly narrowed, with hairs and connected by long membranes. They are more or less retracted in life. Tergite 9 contains two blade-like, triangular, sclerotized plates, but there are no acanthophorites or spines. The whole posterior part can apparently be extended and used as an ovipositor, as in some species of Muscidae. A similar lack of differentiation has been found so far only in the Ethiopian genus *Lepidochlanus*, in which the abdomen is even less differentiated. The structure of the female abdomen has been examined only in a few genera and a similar structure of the abdomen may possibly be found also in other genera. This is a striking example which proves that internal structures may indicate relationship more definitely than external habitus.

In view of the different wing venation and different structure of the female abdomen, it is doubtful whether this group belongs to the subfamily Bombyliinae.

Lordotus Loew, 1863

L. albidus, apiculus, miscellus, pulchrissimus (Figs. 365–370)

Epandrium short, rectangular, with or without projecting posterior corners. Gonocoxites truncate-triangular. Dististyli long, triangular, with narrow, curved apical part. Aedeagus conical, lacking an aedeagal process, with broadly bulging base.

Posterior segments of the female abdomen not differentiated. Tergite 8 normal, visible externally, with hairs. Tergite 9 with two strongly sclerotized, triangular plates. Acanthophorites and spines absent. Spermathecae with more or less long and wide, tubular, pointed capsules. Ducts short, thin. Ejection apparatus short, with short processes and small end plates. Furca V-shaped, with more or less pointed apex.

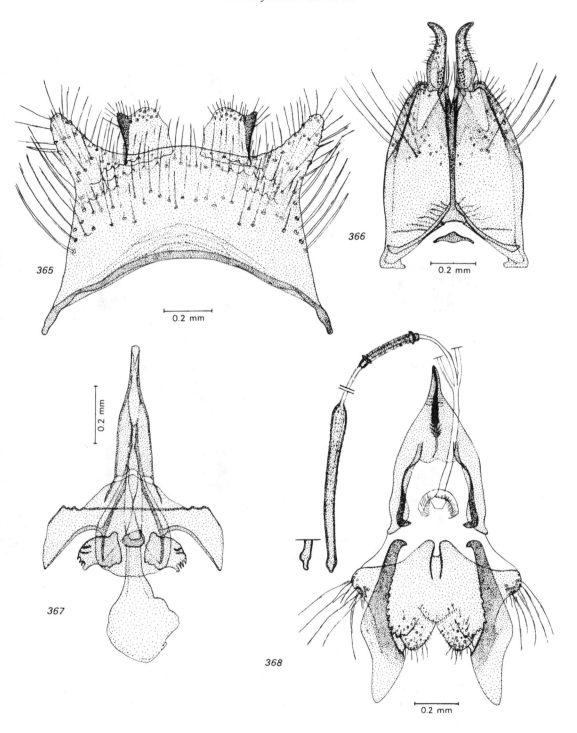

Figs. 365–368: *Lordotus pulchrissimus*
365. epandrium; 366. gonopods; 367. aedeagus; 368. spermatheca and tergite 9

131

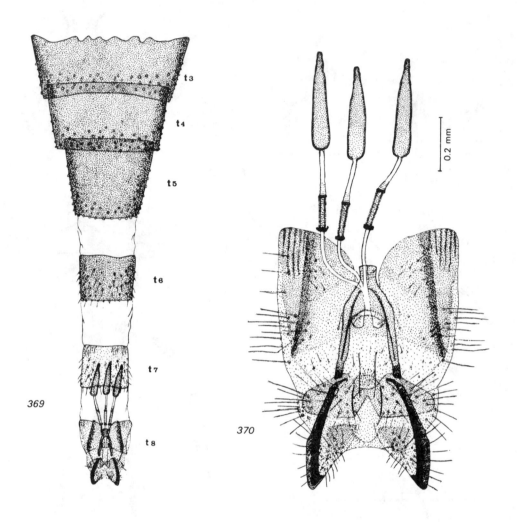

Figs. 369–370:   *Lordotus miscellus*
369. abdomen of female, extended; 370. spermathecae and tergite 9

Geminaria canalis Coquillett, 1894

(Figs. 371–374)

Epandrium short, with broadly truncate, posterior lateral processes. Gonocoxites broadly rounded apically. Dististyli short, triangular, curved. Aedeagus conical; aedeagal process absent.

Abdomen of female as in *Lordotus*. Spermathecae also similar, but capsules with a short, recurved apical process. Ejection apparatus very short, with large crenellated end plates bearing tubercles. Furca V-shaped, with broadly rounded apex. Tergite 9 as in *Lordotus*.

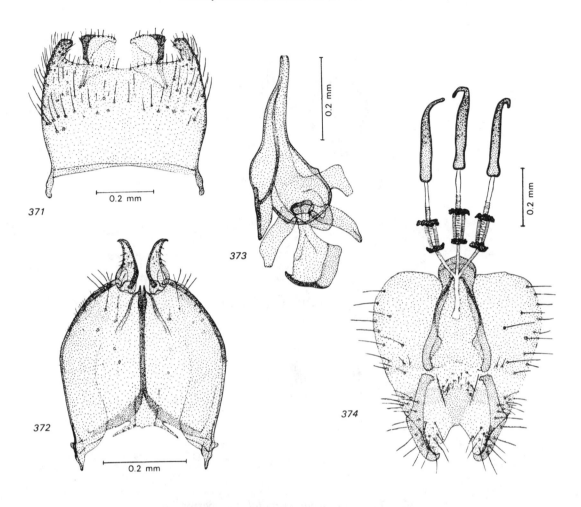

Figs. 371–374: *Geminaria canalis*
371. epandrium; 372. gonopods; 373. aedeagus; 374. spermathecae and tergite 9

Eusurbus crassilabris (Macquart, 1855)

Only a female examined (Fig. 375)

Male hypopygium large, conical, directed downwards, with a tuft of long, bristly hairs on each side according to Hull (1973). Genitalia not described.

Spermathecae with globular capsules with a short apical process with a deep indentation. There is a membranous widening behind the capsule. Ducts wide, short. Ejection apparatus very small, without processes or end plates. Furca with two bars which are curved and widened posteriorly. Tergite 8 completely covered with long hairs, with a long apodeme with a vertical ridge and wide triangular base. Acanthophorites with 27–30 moderately long setae with curved, pointed apex.

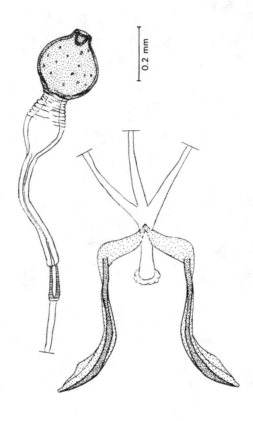

Fig. 375: *Eusurbus crassilabris*, spermatheca

Bombylodes multisetosus (Loew, 1857)
(Figs. 376–378)

Epandrium short, trapezoidal, curved, with small, pointed posterior corners, rounded posterior margin and concave anterior margin. Gonocoxites truncate-triangular, with a small inner apical process. Dististyli triangular, slightly curved. Aedeagus conical, narrow. Aedeagal process with a wide, laterally compressed keel with two small apical processes.

Spermatheca with globular capsules with conical base and short sclerotized ducts. Ejection apparatus long, narrow, striated, without processes and with small end plates. Furca with two bars with widened posterior ends. Tergite 8 with a long apodeme with a wide, triangular base. Acanthophorites with 9–10 short, thick spines.

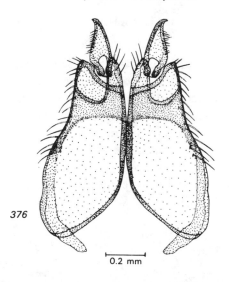

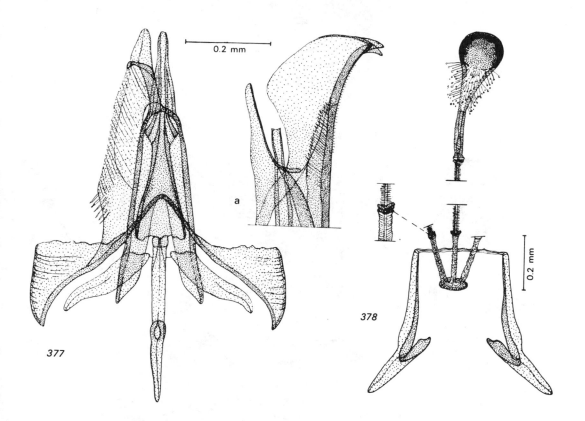

Figs. 376–378: *Bombylodes multisetosus*
376. gonopods; 377. aedeagus; (a) apex of aedeagus, lateral; 378. spermatheca

Acrophthalmida paulseni Philippi, 1865
(Figs. 379–382)

Epandrium trapezoidal, wider posteriorly, with rounded corners and a deep indentation in the middle of the posterior margin. Gonocoxites truncate-triangular, with wider, rounded basal part. Dististyli nearly rectangular, with an apical lateral point. Aedeagus conical; aedeagal process long, club-shaped, with dense, short hairs to near the apex.

Spermathecae with long, broadly club-shaped capsules which pass gradually into thin membranous ducts. Ejection apparatus moderately long, with short and longer processes and a small basal end plate. Only longer processes at the apical end. Ducts to vagina long, very thin. Furca with triangular bars. Tergite 8 with a broadly triangular apodeme. Acanthophorites with 10–12 long spines with curved, pointed apex.

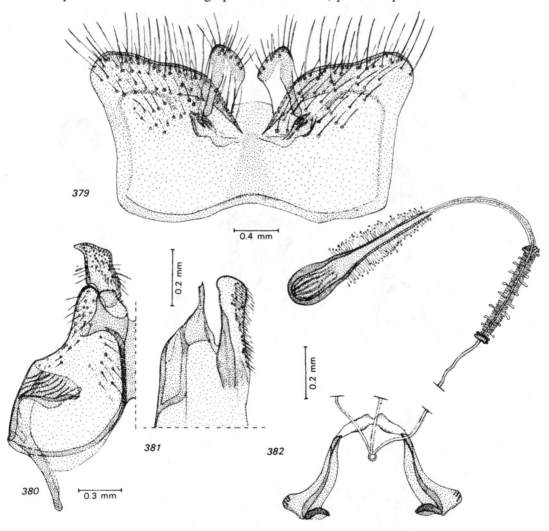

379. 0.4 mm

0.2 mm

0.2 mm

381   382

380   0.3 mm

Figs. 379–382: *Acrophthalmida paulseni*
379. epandrium; 380. gonopod; 381. apex of aedeagus, lateral; 382. spermatheca

Conophorus Meigen, 1803

C. glaucescens, greeni (Figs. 383–386)

Epandrium short, rectangular, curved, with a small, deep indentation in the middle of the posterior margin and with narrow, basal, lateral processes. Gonocoxites broadly

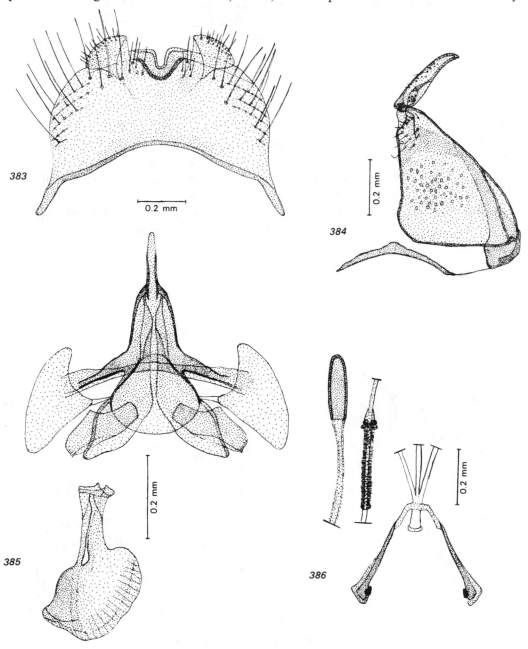

Figs. 383–386 : *Conophorus glaucescens*
383. epandrium; 384. gonopods; 385. aedeagus; 386 spermatheca

triangular; dististyli long, narrow, triangular, slightly curved. Hypandrium very short, crescent-shaped. Aedeagus conical, with wide base and long, narrow apical part. Sheath with lateral shoulders. Aedeagal process absent.

Spermathecae with more or less wide, ovoid or oblong capsules. Ducts very long, relatively wide, membranous, narrowing before the ejection apparatus which is short, narrow, with small processes and a small apical end plate. Furca V-shaped, the two bars with widened ends. Tergite 8 with a long, narrow apodeme having a broad, triangular base. Acanthophorites with about 20 dense spines with curved, pointed ends.

Prorachthes Loew, 1868

Two unidentified species (Figs. 387–389)

Epandrium short, rectangular. Gonocoxites truncate-triangular or nearly parallel-sided, with a narrow apical inner process covered with hairs. Hypandrium very small, crescent-shaped. Dististyli broad basally, with narrow, pointed apical part or triangular. Aedeagus conical; aedeagal process absent.

Capsules of spermathecae ovoid; ducts short, relatively wide, without differentiations, i.e., ejection apparatus apparently absent. Furca with two bars the posterior ends of which are triangularly widened. Tergite 8 with a long, narrow apodeme having a triangular base. Acanthoporites with 2–3 dense rows of about 20 black spines with slightly curved ends.

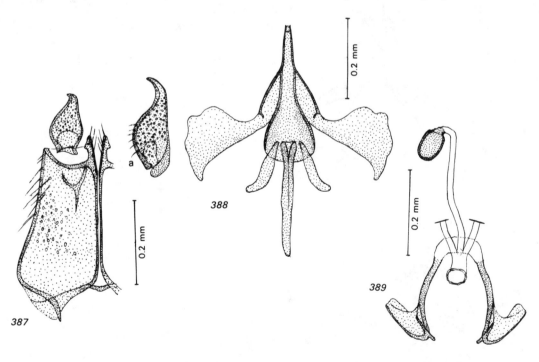

Figs. 387–389: *Prorachthes* sp.
387. gonopods of sp. no. 1; (a) dististylus of sp. no. 2; 388. aedeagus of sp. no. 1;
389. spermatheca of sp. no. 2

Division *TOMOPHTHALMAE Bezzi*

This division is mainly characterized by the structure of the head. The occiput is more or less deeply invaginated and there are two occipital foramina. [See detailed description in the Classification (pp. 13–14).] The other characters on which this division was originally based (indentation of the hind margin of the eyes, a bisection line and the wing venation) are also present in a number of genera of the Homoeophthalmae. The Cythereinae and the *Corsomyza* group (Corsomyzinae) are therefore placed here in the Tomophthalmae.

CYLLENIINAE Becker, 1912

Cyllenia Latreille, 1802

C. maculata and several undetermined species (Figs. 390–396)

The genus *Sphenoidoptera* Williston, 1901 was considered by Painter & Painter (1962) as probably a synonym of *Cyllenia*, and this was repeated by Hall (1969). However, the male genitalia of *Sphenoidoptera* illustrated by Hull (1973) are so different from those described here for *Cyllenia* that this seems unjustified. The genus *Sphenoidoptera* should therefore be maintained and the record of *Cyllenia* from America is apparently incorrect.

Epandrium short, more or less rectangular, concave posteriorly, with lateral posterior processes of specific form which bear dense, short spines in some species. Cerci with short, dense, black spines. Gonocoxites truncate-triangular, with inner apical processes. Hypandrium large, triangular. Dististyli long, narrow, pointed. Aedeagus long, narrow, curved at the base, so that the apodeme is situated at a right angle to the long axis of the aedeagus or even directed obliquely posteriorly. Aedeagus divided into three prongs in the apical half, the median prong being longer than the lateral ones in one species. Aedeagal process long, V-shaped, with long, curved apical part. Base of sheath also with two long lateral processes having a pointed apex and sparse short hairs.

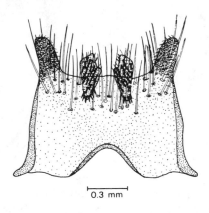

0.3 mm

Fig. 390: *Cyllenia maculata* (Europe), epandrium

139

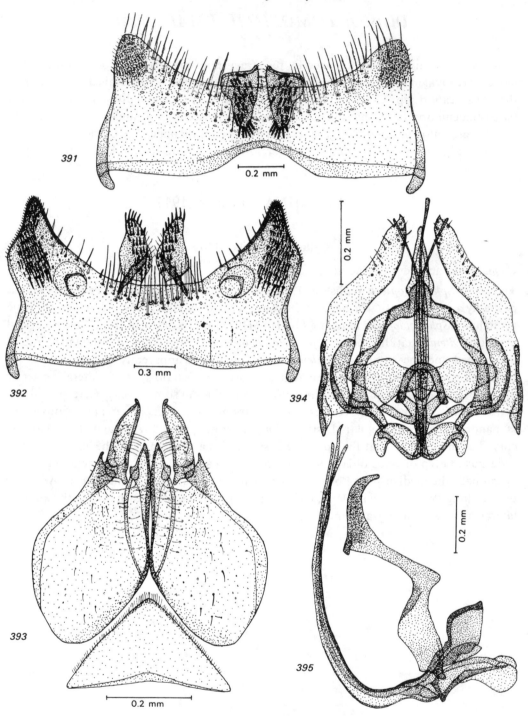

Figs. 391–395: *Cyllenia* sp. no. 1
391. epandrium; 392. same, sp. no. 2; 393. gonopods of sp. no. 2;
394. male genitalia, flat; 395. aedeagus

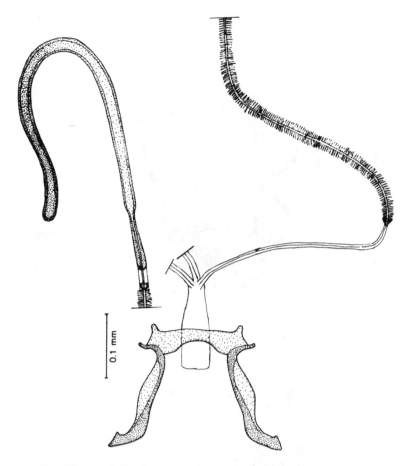

Fig. 396 : *Cyllenia* sp. no. 1, spermatheca

Spermathecae with long, tubular, recurved capsules which narrow proximally before the ejection apparatus. Ducts very short. Ejection apparatus very long, narrow, with small processes and small end plates. Ducts to vagina long, narrow; common duct wide, membranous. Furca rectangularly U-shaped. Tergite 9 divided, spines absent.

Amictus Wiedemann, 1817

A. minor, obliquenotatus, setosus, tigrinus, validus, virgatus (Figs. 397–404)

Epandrium varying markedly in form, size and chaetotaxy in the different species. It is large, concave posteriorly, convex anteriorly, with dense, long setae in *validus*, so that the posterior end of the abdomen is markedly wider than the other abdominal segments, or smaller, rectangular, with shorter airs (*minor, setosus, tigrinus*). There is a triangular or rounded process between the cerci, which apparently belongs to the proctiger in some species (*minor, setosus*). Cerci with dense, short, black spines (*validus, tigrinus, virgatus*) or with more numerous thin spines in *minor*. Gonocoxites truncate-triangular; dististyli long,

141

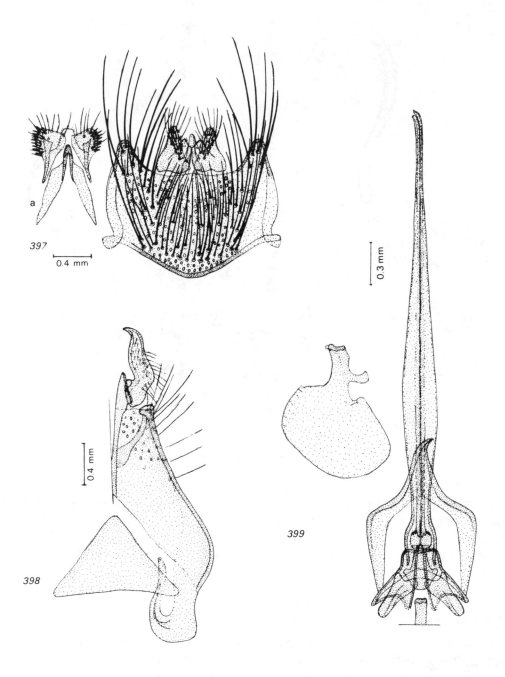

Figs. 397–399:   *Amictus validus*
397. epandrium; (a) cerci, different aspect; 398. gonopod; 399. aedeagus

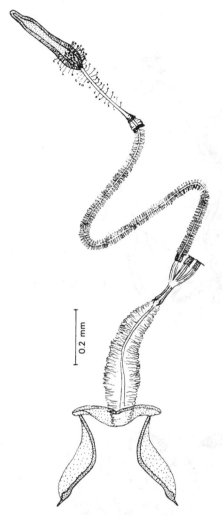

Fig. 400: *Amictus validus,* spermatheca

triangular, with sinuate sides, pointed apex and specific differences. Hypandrium large, triangular. Aedeagus very long, thin, with recurved base, so that the apodeme is situated at a right angle to the long axis of the aedeagus or directed obliquely posteriorly. Apical part of aedeagus with wide lateral extensions in *validus.* Aedeagal process V-shaped, with long, curved, pointed apical part.

There are three distinctly different types of spermathecae in this genus:

Type 1. Capsules short, tubular, pointed. Ducts as long as the capsule, very thin. Ejection apparatus very long, with short processes and small end plates. Common duct very wide, membranous, striated (*validus, virgatus*).

Type 2. Capsules forming very long, sclerotized tubes which become very narrow before the ejection apparatus. Ducts very short. Ejection apparatus very long, without end plates (*obliquenotatus*).

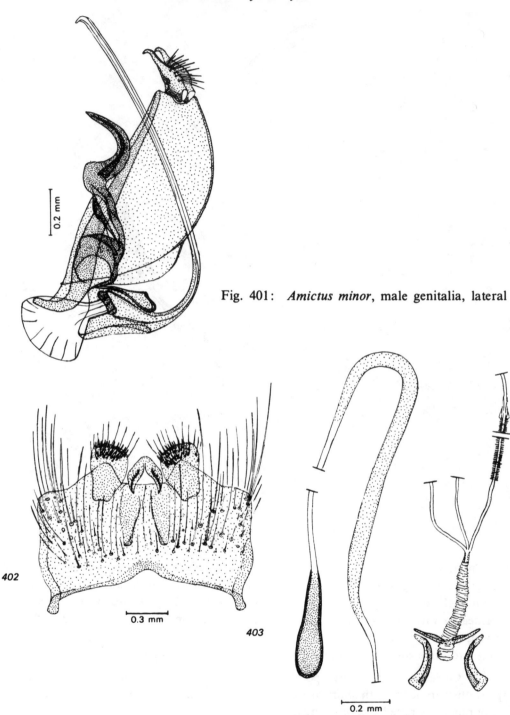

Fig. 401: *Amictus minor*, male genitalia, lateral

Figs. 402–403: *Amictus setosus*
402. epandrium; 403. spermatheca

Type 3. Capsules oblong-ovoid, pear-shaped or club-shaped. Ducts extremely long, widening before the ejection apparatus which is also very long and narrow. Common duct wide, striated (*setosus, minor, tigrinus*). Furca rectangularly U-shaped; bars with specific differences. Tergite 8 with a laterally compressed apodeme with a triangular base. Acanthophorites with 3–6 long spines.

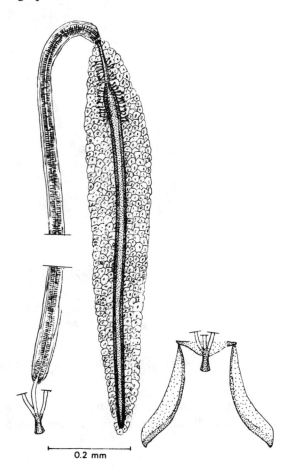

0.2 mm

Fig. 404: *Amictus obliquenotatus*, spermatheca

Sinaia kneuckeri Becker, 1916
(Fig. 405)

This genus contains only one species. The second species, *S. aharoni* Engel, proved to be a synonym of *Amictus minor* Austen.

The male genitalia differ only in minor details from those of *Amictus*. Epandrium nearly rectangular, with sparse, short hairs. Cerci with numerous short spines.

Spermathecae of type 3 of *Amictus*, but capsules very narrow, club-shaped, with rounded

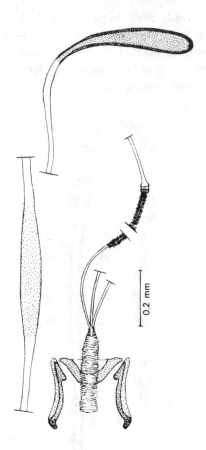

Fig. 405: *Sinaia kneuckeri*, spermatheca

apex. Ducts and ejection apparatus very long, thin; ducts widened in the middle. Furca with two bars and a broadly angular sclerite between their apical ends. Acanthophorites with 6–7 long spines.

Paracosmus Osten-Sacken, 1877

P. edwardsi, rubicundus (Figs. 406–413)

P. rubicundus. Epandrium short, rectangular, with broadly rounded projecting posterior corners. Gonocoxites fused; dististyli long, parallel-sided, with an apical lateral point. Hypandrium absent. Aedeagus conical; aedeagal process absent. Apodeme large, rounded-triangular, with processes near the head.
Spermathecae with ovoid-oblong capsules. Ducts very long, sclerotized in their greater part, narrowing proximally. Ejection apparatus short, with wide end plates and very short processes. Furca with two bars with triangularly widened base. Tergite 8 very short, with a long, laterally compressed apodeme having a triangular base. Acanthophorites with four long spines.

146

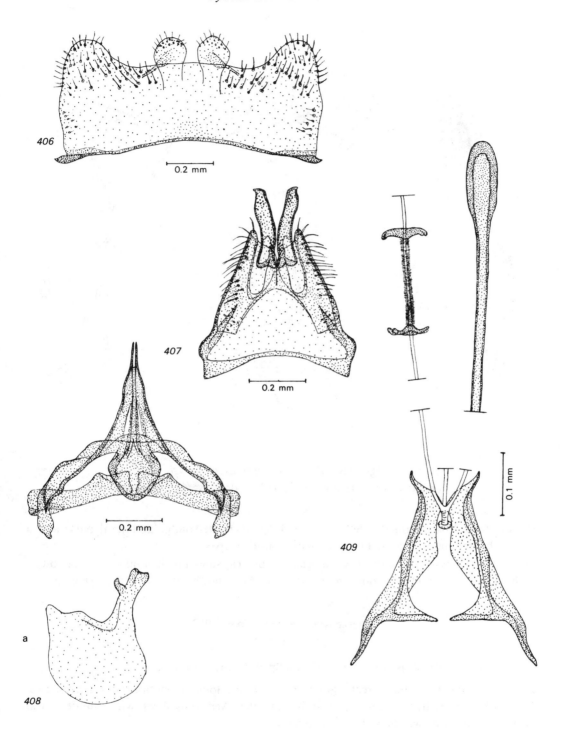

Figs. 406–409: *Paracosmus rubicundus*
406. epandrium; 407. gonopods; 408. aedeagus; (a) apodeme; 409. spermatheca

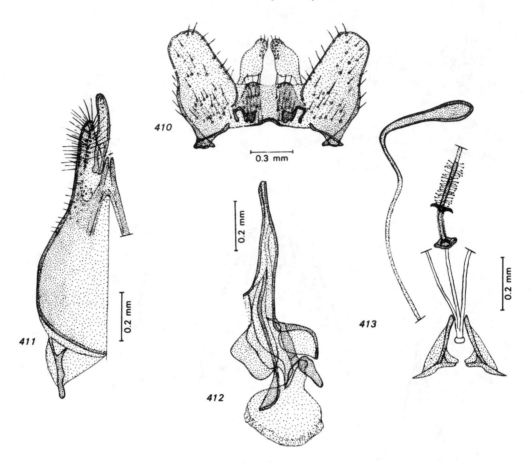

Figs. 410–413: *Paracosmus edwardsi*
410. epandrium; 411. gonopod; 412. aedeagus; 413. spermatheca

P. edwardsi. Epandrium divided into two long, truncate-triangular lateral parts and a much shorter median part. Dististyli with rounded apex.

Spermathecae resembling those of *rubicundus*, but capsules much smaller and narrower, club-shaped. Ejection apparatus much shorter. Acanthophorites with two long spines.

Metacosmus nitidus Cole, 1923

(Figs. 414–419)

Two males identified as *nitidus* had distinctly different genitalia.

Type a. Epandrium short, rectangular, with broad, long, posterior lateral processes. Gonocoxites broadly triangular. Dististyli triangular. Aedeagus short, with rounded basal part. Sheath with two curved, apical points.

Type b. Epandrium longer, rectangular, with broadly rounded posterior corners. Gonocoxites truncate-triangular. Dististyli oblong, with an apical point and a vertical

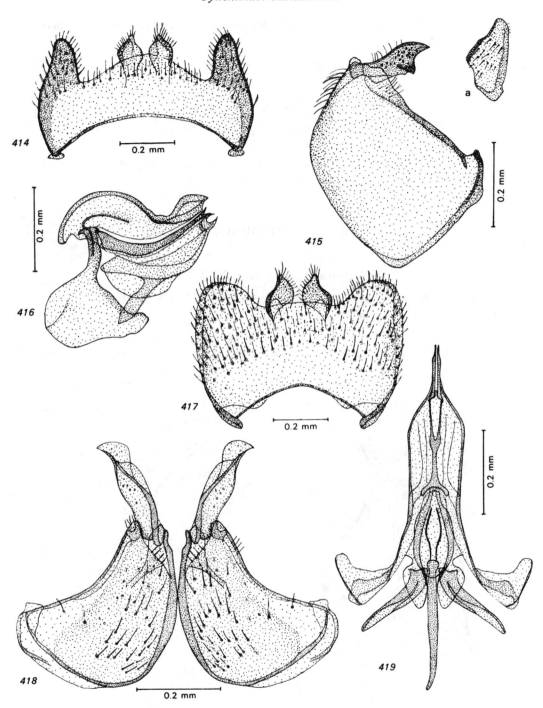

Figs. 414–419: *Metacosmus nitidus*
414. epandrium, type a ; 415. gonopod, type a ; (a) dististylus, lateral ;
416. aedeagus, type a; 417. epandrium, type b; 418. gonopods, type b; 419. aedeagus, type b

ridge. Aedeagus long, parallel-sided in its greater basal part, tapering to a thin point in the apical quarter.

It is not clear which of the two types is *nitidus.*

Amphicosmus elegans Coquillett, 1891

A female examined has spermathecae resembling those of *Paracosmus rubicundus*, but the capsules are larger and more oblong-oval. Ejection apparatus with small end plates, the apical plate with an apical process. Furca V-shaped; bars triangular, with posterior processes. Acanthophorites with four long spines.

Corsomyza Group (CORSOMYZINAE)

Mariobezzia Becker, 1912

M. lichtwardi, catherinae, and an undetermined species (Figs. 420–424)

Epandrium short, rectangular, with rounded posterior corners and a small, deep indentation in the middle of the posterior margin. Cerci triangular, with a dark sclerotization in *catherinae.* Gonocoxites truncate-triangular, slightly curved, with specific

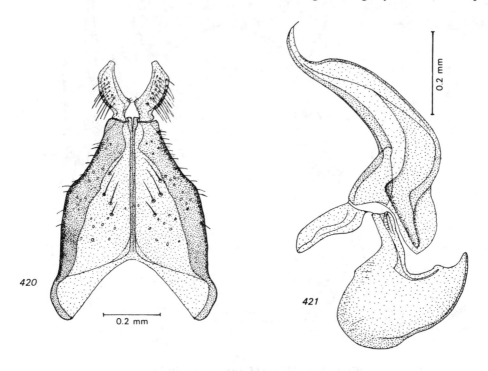

Figs. 420–421: *Mariobezzia lichtwardi*
420. gonopods; 421. aedeagus

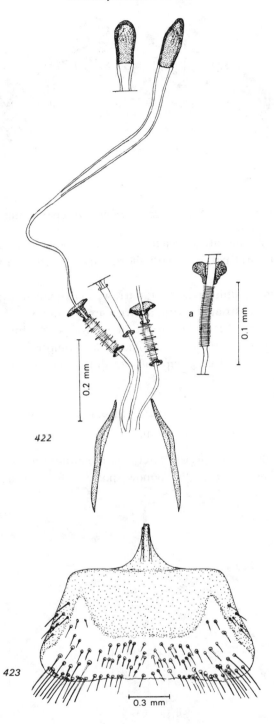

Figs. 422–423 : *Mariobezzia lichtwardi*
422. spermatheca ; (a) other species ; 423. tergite 8 of female

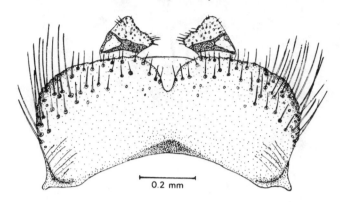

Fig. 424: *Mariobezzia catherinae*, epandrium

differences. Hypandrium absent. Aedeagus simple, without aedeagal process, slightly tapering in its greater basal part and with short, curved, narrow apical part. Apodeme triangular.

Spermathecae with small, oblong-oval or narrow, club-shaped capsules. Ducts long. Ejection apparatus short, striated, with a large apical and small basal end plate in *lichtwardi*, with only an apical end plate in the other two species. Furca with two separate, thin bars. Tergite 8 long, with a narrow, laterally compressed apodeme. Tergite 9 transversely rectangular, with two groups of 4–5 short, thick spines with widened apex. Acanthophorites absent.

Corsomyza brevicornis Hesse, 1938 (Fig. 425), **Megapalpus capensis** Wiedemann, 1828,
Zyxmyia megachile Bowden, 1960

The male genitalia of the three species closely resemble those of *Mariobezzia,* except for minor differences. Epandrium broadly rounded posteriorly, with an indentation in the

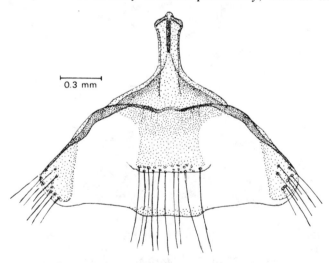

Fig. 425: *Corsomyza brevicornis*, tergite 8 of female

middle of the posterior margin. The apical part of the aedeagus is more or less shortened and not curved. The spermathecae of *Corsomyza* have larger, ovoid capsules, while those of *Zyxmyia* are tubular, short, with a rounded apex. Tergite 8 of the three species is very characteristic, having a broad apodeme, and is more or less triangular with a distinct pattern of pigmentation. It differs distinctly from that of *Mariobezzia*. Tergite 9 with 5–7 large, curved spines. Acanthophorites not separated. Spines situated on a tubercle in *Zyxmyia*.

CYTHEREINAE Becker, 1912

This subfamily was considered as a tribe of the Bombyliinae by Hull (1973). This is incorrect. It belongs to the Tomophthalmae and its rank as a subfamily seems justified.

Cytherea Fabricius, 1794

C. albolineata, aurea, beckeri, (?) *bisalbifrons, carmelitensis, delicata, dispar,* (?) *maroccana, nucleorum, obscura, syriaca, thyridophora,* and several undescribed species (Figs. 426–436)

Epandrium nearly rectangular or trapezoidal, wider proximally. Gonocoxites truncate-triangular. Dististyli more or less narrow, triangular or with distinctly wider basal part and narrow, curved apical part.

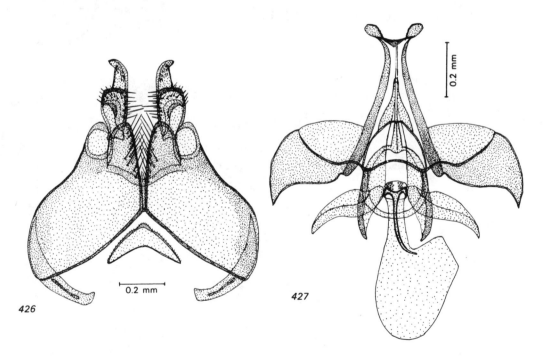

Figs. 426–427: *Cytherea delicata*
426. gonopods; 427. aedeagus

153

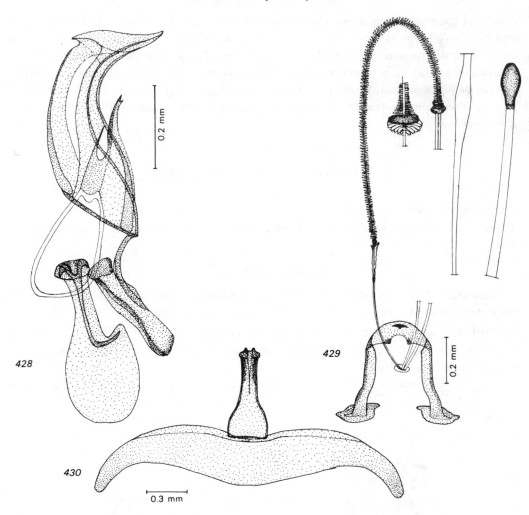

Figs. 428–430: *Cytherea delicata*
428. aedeagus, lateral; 429. spermatheca; 430. tergite 8 of female

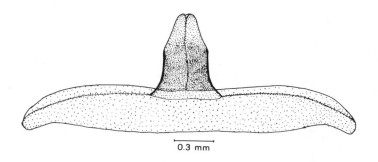

Fig. 431: *Cytherea nucleorum*, tergite 8 of female

154

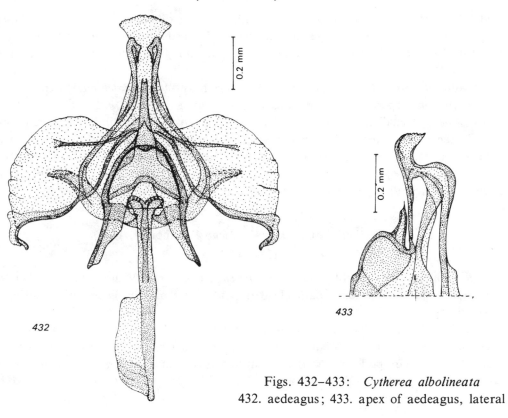

Figs. 432–433: *Cytherea albolineata*
432. aedeagus; 433. apex of aedeagus, lateral

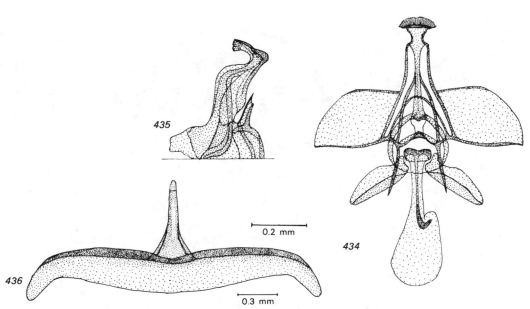

Figs. 434–436: *Cytherea syriaca*
434. aedeagus; 435. apex of adeagus, lateral; 436. tergite 8 of female

Aedeagus conical. Aedeagal process long, with distinctly differentiated apical part, often with a pointed, proximally directed apical process. The aedeagus of *dispar* and *albolineata* has a distinct ventral bulge with strongly sclerotized marginal ridges before the narrow apical part.

Spermathecae with ovoid or oblong, sausage-shaped capsules. Ducts long, widened in the middle in some species. Ejection apparatus very long, narrow, with uniformly small processes and a small apical end plate. The basal sclerotization is cylindrical, more or less long in the various species. The apical end plate shows membranous differentiations in some species. Furca rectangularly U-shaped, with specific differentiations. Tergite 8 with a more or less wide apodeme, very wide in *nucleorum* and (?) *maroccana*, T-shaped in *albolineata*. Acanthophorites with 5–6 spines in some species, 12–18 in others.

Callostoma fascipenne palaestinae Paramonov, 1931
(Fig. 437)

The genitalia closely resemble those of *Cytherea*. Aedeagus very short, broadly conical. Aedeagal process with long, parallel-sided apical part and widened apex bearing a triangular process.

Spermathecae with small ovoid capsules. Ducts widened behind the capsule, much shorter than in *Cytherea*. Furca U-shaped with widened basal ends. Tergite 8 with a wide, truncate-triangular apodeme. Ejection apparatus also as in *Cytherea*, but much shorter. Acanthophorites with 9–10 long, thin spines.

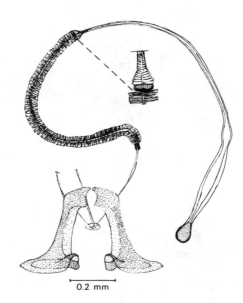

0.2 mm

Fig. 437: *Callostoma fascipenne palaestinae*, spermatheca

Pantarbes pusio Osten-Sacken, 1877
(Figs. 438–440)

Epandrium short, rectangular; cerci club-shaped. Gonocoxites truncate-triangular; dististyli long, narrow, narrower in the middle, apex widened, rounded, with a short, curved, pointed process. Aedeagus conical; aedeagal process not reaching apex of aedeagus. Apodeme large, rounded-triangular. The drawing of the genitalia of another species given by Hull is quite different.

Spermathecae of the same type as in *Cytherea*, but much shorter. Capsules ovoid, pointed; ducts widened behind the capsule. Ejection apparatus not longer than the ducts, with only small processes and indistinct end plates. Furca with two bars with inner basal processes and with a triangular sclerite between their apical ends. Tergite 8 with a large triangular apodeme with a vertical ridge. Acanthophorites with 2–3 long, thin spines.

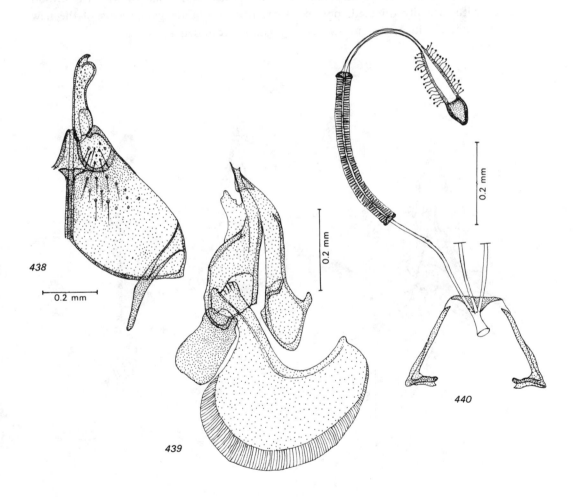

Figs. 438–440: *Pantarbes pusio*
438. gonopod; 439. aedeagus; 440. spermatheca

LOMATIINAE Schiner, 1868

Plesiocera Macquart, 1840

P. algira, europaea, nigrirostris, and an undescribed species (Figs. 441–454)

Stomylomyia Bigot is here considered as a subgenus of *Plesiocera* following Engel (1935). Epandrium trapezoidal, with specific differences in form. Gonocoxites more or less wide, triangular, with a pointed apical lateral process in *algira*. Dististyli of distinctly different form in the various species, either widening apically and with a pointed apical process (*algira*), or having a wide base with a lateral angle and a more or less long apical process. Hypandrium triangular, of varying form and size. Aedeagus conical. Aedeagal process of varying form, V-shaped with more or less long, curved, pointed apex (*europaea, nigrirostris*), as long as or slightly shorter than the aedeagus. Sheath with two broad processes with laterally directed, pointed apex in *algira*. Aedeagal process of the new species V-shaped, with wide, thin-walled, irregularly rounded apex.

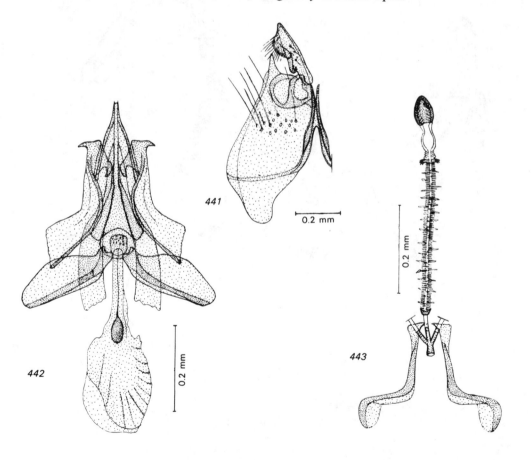

Figs. 441–443: *Plesiocera algira*
441. gonopod; 442. aedeagus; 443. spermatheca

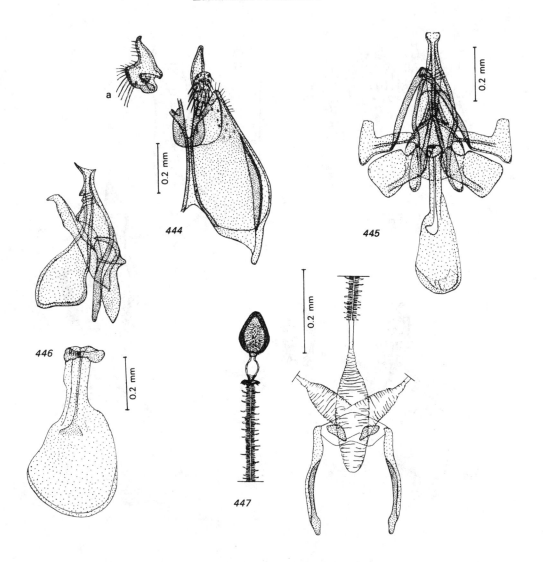

Figs. 444–447: *Plesiocera nigrirostris*

444. gonopod; (a) dististylus, lateral; 445. aedeagus; 446. same, lateral; 447. spermatheca

The spermathecae of *algira* and *nigrirostris* have small, pointed, thick-walled capsules. Ducts very short, widened, membranous. Ejection apparatus long, with short and longer processes and a distinct apical end plate. Basal end plate small or absent. Ducts to vagina short and narrow in *algira*, but very wide, membranous and striated in *nigrirostris*. Furca U-shaped, narrower apically and with laterally curved basal parts in *algira*. Tergite 8 with a long, narrow apodeme in *algira*, with a triangular apodeme with a vertical ridge in *nigrirostris*. Acanthophorites with 6–7 very long, thin spines in *algira*, with 5–6 short, thick spines in *nigrirostris*.

159

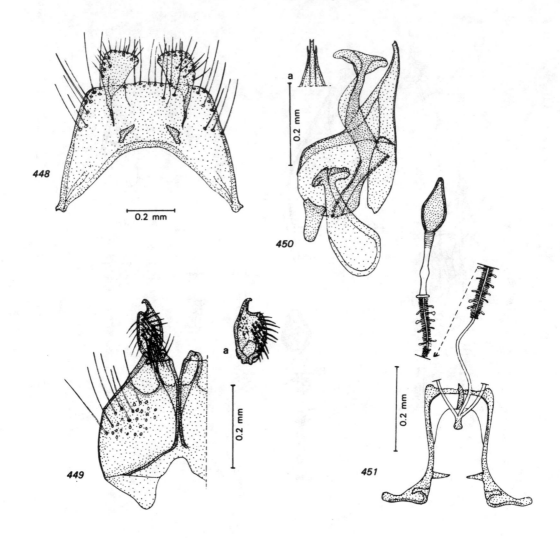

Figs. 448–451: *Plesiocera* sp. no. 1
448. epandrium; 449. gonopod; (a) dististylus, lateral; 450. aedeagus;
(a) same, apex, dorsal; 451. spermatheca

The spermathecae of the new species are of the same type, but the capsules are larger, less strongly sclerotized, rhomboidal, and the ducts and ejection apparatus are longer and narrower.

The spermathecae of *P. (Stomylomyia) europaea* are distinctly different. The capsules are much larger, oblong-oval, with rounded apex. The ducts are very long. Ejection apparatus very narrow, with small and a few longer processes and small end plates. Furca U-shaped; ducts to vagina very thin and long. Tergite 8 with a long apodeme with a triangular base. Acanthophorites with five thick spines.

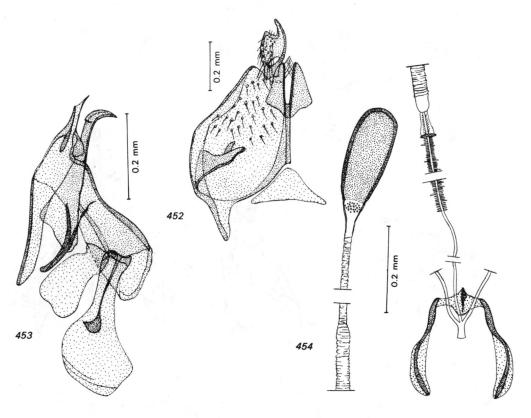

Figs. 452–454: *Plesiocera (Stomylomyia) europaea*
452. gonopod; 453. aedeagus, lateral; 454. spermatheca

Aphoebantus Loew, 1872

A. efflatouni, stenurus (Palaearctic); *A. conurus, eremicola, fumosus, obtectus, rattus, varius* (Nearctic) (Figs. 455–474)

The genus was based on the Nearctic species *cervinus* Loew. The Palaearctic species differ distinctly from the Nearctic species examined in several characters and may have to be placed in a subgenus.

A. efflatouni, stenurus. Epandrium rectangular, with slightly projecting, rounded posterior corners and concave proximal margin. Cerci with numerous thin hairs. Gonocoxites broad, triangular, with a lateral rounded process in *stenurus*. Dististyli triangular, with narrow curved, apical part. Hypandrium absent. Aedeagus long, thin, slightly S-curved. Aedeagal process not reaching end of aedeagus, with bifid apex and two large plate-shaped processes in the middle in *stenurus*, with pointed, curved apex, not bifid in *efflatouni*. Spermathecae with rounded or slightly oblong-oval capsules, continuing in a sclerotized, tapering duct which then becomes thin and membranous. Ejection apparatus as long as ducts, with short and longer processes and end plates with processes. Ducts to vagina

161

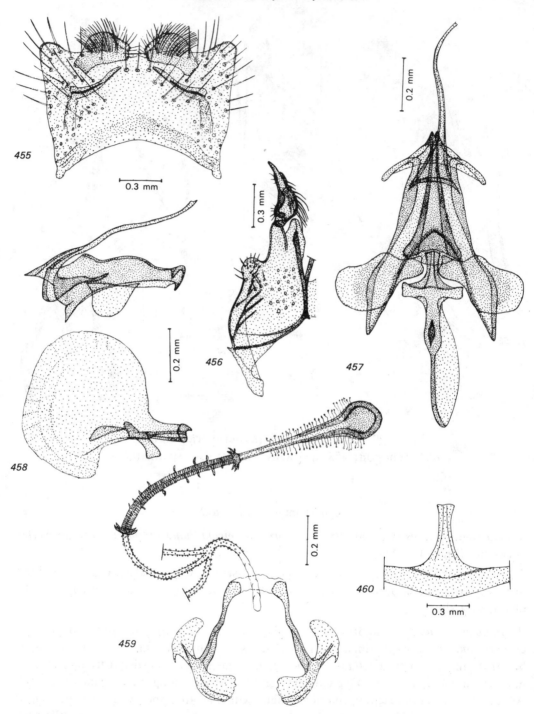

Figs. 455–460: *Aphoebantus stenurus*

455. epandrium; 456. gonopod; 457. aedeagus; 458. same, lateral; 459. spermatheca; 460. tergite 8, apodeme

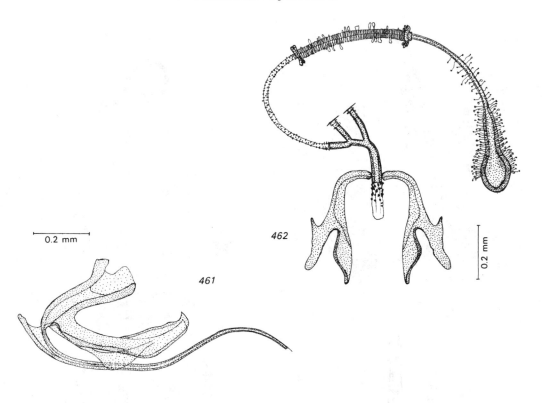

Figs. 461–462: *Aphoebantus efflatouni*
461. aedeagus; 462. spermatheca

sclerotized at the base, ending in a short common duct which is sclerotized in *efflatouni*. Furca broadly U-shaped; bars broadly bifid posteriorly in *efflatouni*, with two broad, transverse, basal processes in *stenurus*. Tergite 8 with a long, narrow apodeme in both species. Acanthophorites with 20–30 long, dense, hair-like setae in *stenurus*, with five thick spines with curved apex in *efflatouni*.

The Nearctic species differ distinctly from the Palaearctic species in several characters (form of aedeagus, form of tergite 8 in the female and form of the spermathecae). Some of the Nearctic species (*rattus*) also have a projecting face so that this can apparently not be used as a generic character in every case.

Epandrium and gonocoxites resemble those of the Palaearctic species, but there are exceptions, e.g., the gonocoxites of *eremicola* are very wide and rounded and have a curved apical process. The dististyli are of distinctive form in some species, e.g., in *rattus*, in which they have an additional apical tubercle. The aedeagus of the species examined is much shorter than that in the Palaearctic species and the aedeagal process shows distinct specific differentiations.

163

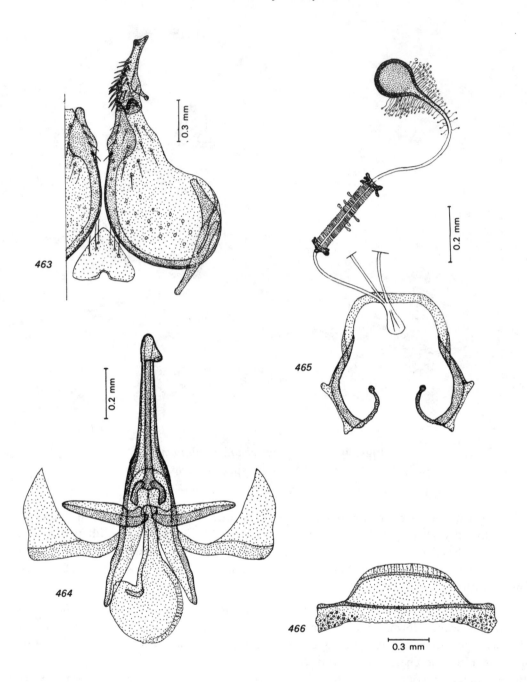

Figs. 463–466: *Aphoebantus rattus*
463. gonopod; 464. aedeagus; 465. spermatheca; 466. tergite 8 of female, apodeme

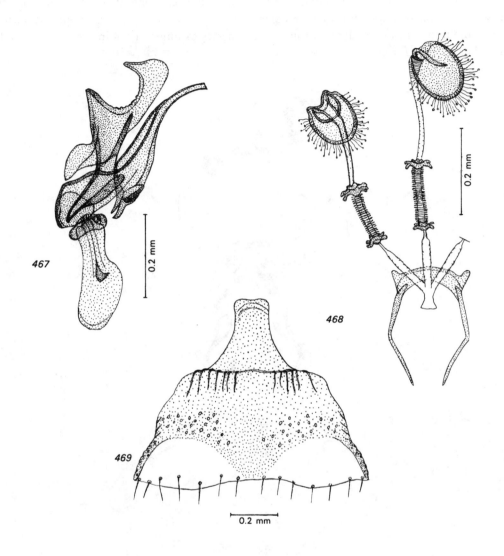

Figs. 467–469: *Aphoebantus obtectus*
467. aedeagus ; 468. spermathecae ; 469. tergite 8 of female

The spermathecae vary markedly in form. In some species (*rattus*) they resemble those of the Palaearctic species, except for the form of the furca. They are very short in *conurus*; the capsules are oblong in *varius*, very large, with pointed apex in *eremicola*, broadly oval and recurved in *obtectus*. Ejection apparatus very short, strongly sclerotized, with large end plates in most species, but very long and narrow in *fumosus*. Furca U-shaped but with distinct specific differences and characteristic adjacent sclerites in most species. There is a

165

distinct difference from the Palaearctic species in the form of tergite 8. The apodeme is very wide, rounded, or truncate-triangular, but of varying size. None of the Nearctic species have a long, narrow apodeme like the Palaearctic species examined. Acanthophorites with a varying number of curved, thick spines in all species examined (3–4 in some species, 6–8 in others, 13 in *rattus*).

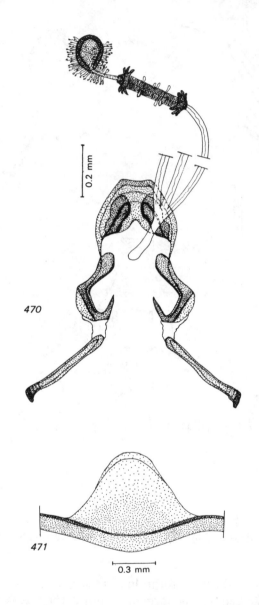

Figs. 470–471: *Aphoebantus conurus*
470. spermatheca; 471. tergite 8 of female, apodeme

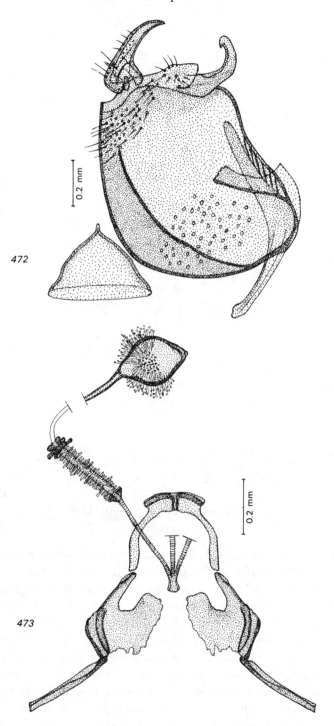

Figs. 472–473: *Aphoebantus eremicola*
472. gonopod; 473. spermatheca

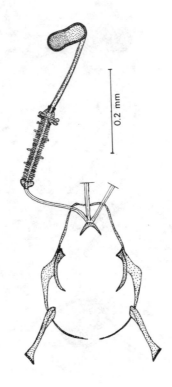

Fig. 474: *Aphoebantus varius*, spermatheca

Epacmus litus Coquillett, 1886
(Figs. 475–479)

Epandrium rectangular, with an indentation in the middle of the posterior margin. Cerci triangular, black, with numerous short, black spines and a V-shaped process between them as in some species of *Amictus*. Gonocoxites broad, triangular, with an apical process with an inward directed, narrow apical process with a broad base. Dististyli long, narrow, tapering, with recurved, pointed apex. Hypandrium large, triangular. Aedeagus conical; aedeagal process long, V-shaped, with curved, tapering apical part.

Spermathecae with short, tubular capsules with rounded apex. Ducts short, thin. Ejection apparatus short, with short processes and small end plates. Furca with two bars and a triangular sclerite between their apical ends. Tergite 8 resembling that of some species of *Aphoebantus*, triangular, with a broad anterior apodeme and a median and two lateral areas of pigmentation. Acanthophorites with 6–8 strong, curved spines.

The genitalia described differ distinctly from the illustrations of a species of *Epacmus* given by Hull (1973).

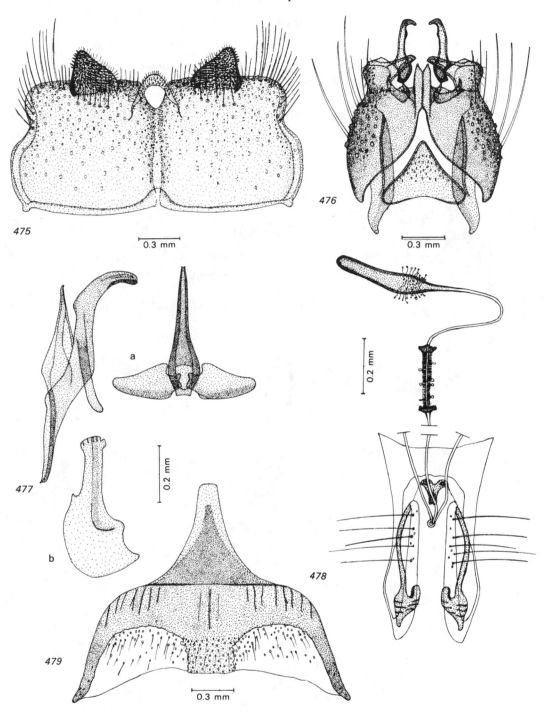

Figs. 475–479: *Epacmus litus*

475. epandrium; 476. gonopods; 477. aedeagus, lateral; (a) same, dorsal; (b) apodeme; 478. spermatheca; 479. tergite 8 of female

Petrorossia Bezzi, 1908

P. hesperus, letho, and two undescribed species (Figs. 3, 480–492)

Most species are very similar in external characters, but examination of the genitalia of both sexes showed distinct differences in specimens not distinguishable by external characters.

Epandrium broadly rounded posteriorly, with lateral basal corners. Gonocoxites very broad, truncate-triangular. Dististyli nearly rectangular, with short processes at the apex,

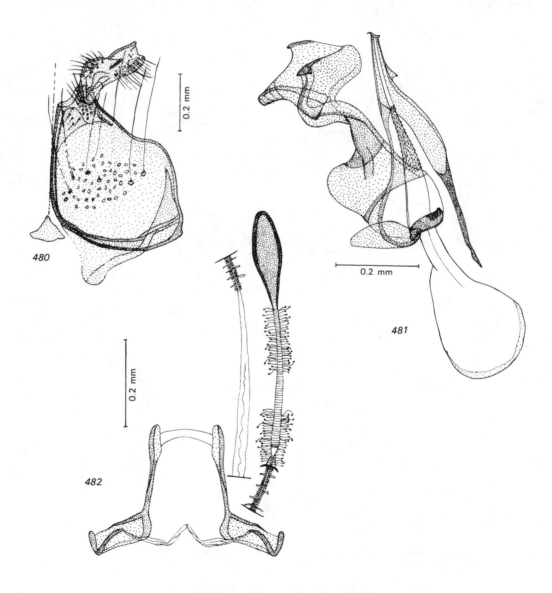

Figs. 480–482 : *Petrorossia hesperus*
480. gonopod ; 481. aedeagus, lateral ; 482. spermatheca, type 1

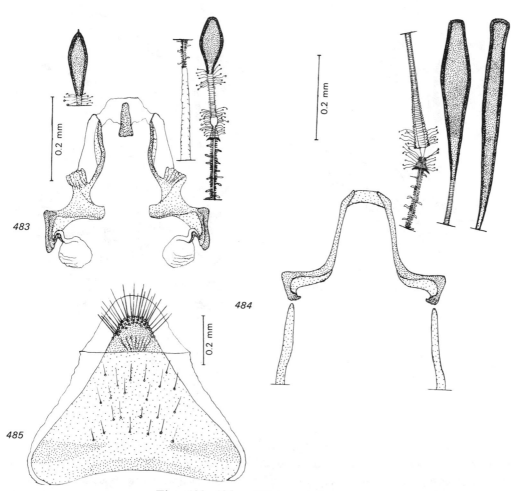

Figs. 483–485: *Petrorossia hesperus*
483. spermatheca, type 2; 484. same, types 3 and 4;
485. type 3, posterior sternite of female

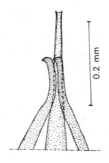

Fig. 486: *Petrorossia letho*, apex of aedeagus

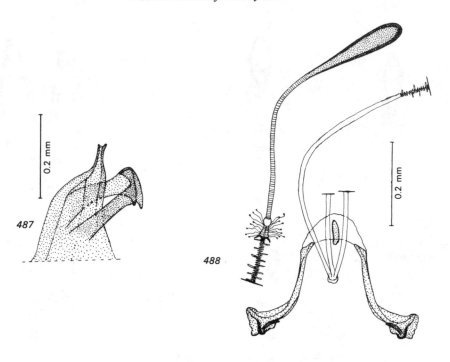

Figs. 487–488: *Petrorossia* sp. no. 1
487. apex of aedeagus, lateral; 488. spermatheca

with specific differences, very wide apically in *letho*. Hypandrium very small, triangular. Aedeagus of *hesperus* conical, with two denticles at the base of the apical part; aedeagal process of complicated form, very wide, with two short apical processes and two large, recurved, pointed lateral processes. Aedeagus of *letho* with long, narrow apical part; aedeagal process simple, V-shaped, with two apical curved points at the apex. Aedeagal process of species no. 1 broad, with two apical lateral points, that of species no. 2 V-shaped, with broadly rounded apex.

Spermathecae of specimens identified as *hesperus* of three or four types.

Type 1. Capsules ovoid, with short, striated ducts. Ejection apparatus narrow, very long, with short and longer processes. Furca with two bars with widened base bearing characteristic processes. Acanthophorites with 2–3 long spines.

Type 2. Capsules much longer. Basal part of furca rectangular, with an S-shaped ridge.

Type 3. Capsules long, narrow, club-shaped and with long thin ducts. Basal part of furca narrow, rectangularly bent. Acanthophorites with 5–6 long spines. One specimen had sperm capsules of different form and may belong to yet another type.

The two new species have similar spermathecae with specific differences. Species no. 1 apparently has no spines on tergite 9.

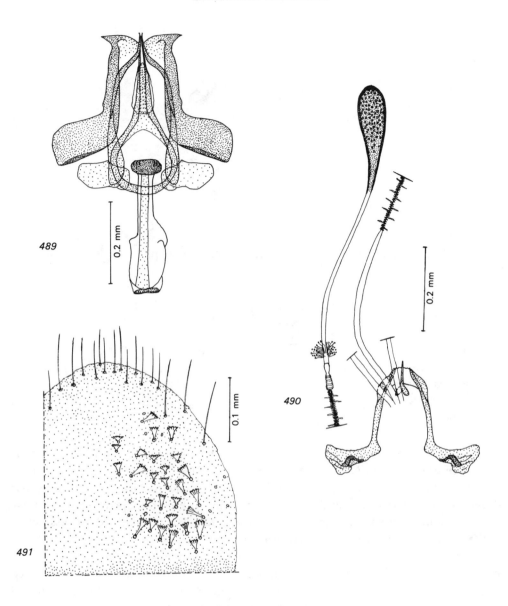

Figs. 489–491: *Petrorossia* sp. no. 2
489. aedeagus; 490. spermatheca; 491. scales on posterior sternite of female

A similar situation was found in *letho*, with two different types of sperm capsules. Acanthophorites with 5–6 long, thin spines.

The spermathecae show some variation of form in the same species, but the differentiations of the basal part of the furca are constant in every form.

Tergite 8 of all species examined with a long, narrow apodeme. Further variation of the male genitalia was illustrated by Hesse (1938).

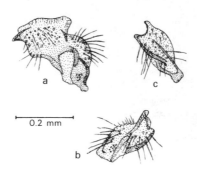

Fig. 492: *Petrorossia* spp., dististylus; (a) *P. letho*; (b) sp. no. 1; (c) sp. no. 2

Pipunculopsis Bezzi, 1925

P. bivittata and a new species (Figs. 493–499)

This genus was considered as a subgenus of *Petrorossia* by Engel (1936), but it differs so distinctly from this in external characters and in the genitalia that it seems justified to maintain it as a genus.

Epandrium relatively long, trapezoidal, with projecting posterior corners in *bivittata*, with rounded corners in the new species. Cerci with a lateral bulge and a rounded apical process in *bivittata*, of different form in the new species. Gonocoxites truncate-triangular. Dististyli with broadly rounded base and long, narrow, curved apical process bearing a subapical denticle in *P. bivittata*, with nearly rectangular basal part bearing strong, short spines and a short apical process in the new species. Hypandrium large, triangular. Aedeagus short, conical; aedeagal process very long, with club-shaped apex and dense, short spines and short hairs at the apex. The new species has no aedeagal process but the sheath bears 10 long, slender processes which are longer laterally.

Spermathecae of *bivittata* with tubular capsules which are widened at the base. Ducts short, narrow. Ejection apparatus much longer than ducts, with short and longer processes. Apical end plate cup-shaped; basal end plate short, cylindrical. Furca U-shaped, with long processes at the base and an inner apical process. Tergite 8 with a long, narrow apodeme.

The spermathecae of the new species have small, conical, pigmented capsules continuing in a wide membranous duct which narrows proximally. Ejection apparatus very long, with short and long processes and a small apical end plate. Furca consisting of three separate sclerites, the proximal sclerites wide, with rounded processes, the apical sclerite large, T-shaped, with a long posterior process. Tergite 8 as in *bivittata*. Acanthophorites with one seta.

The two species are very similar in habitus and coloration, but the new species has a distinctly projecting face.

174

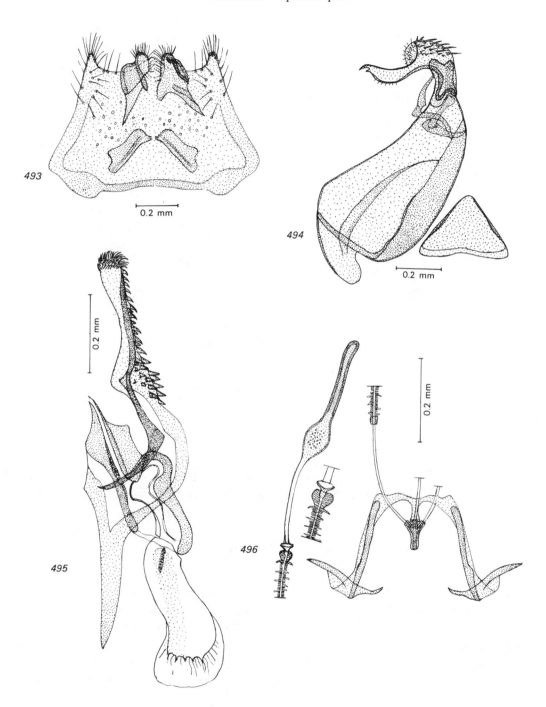

Figs. 493–496: *Pipunculopsis bivittata*
493. epandrium; 494. gonopod; 495. aedeagus; 496. spermatheca

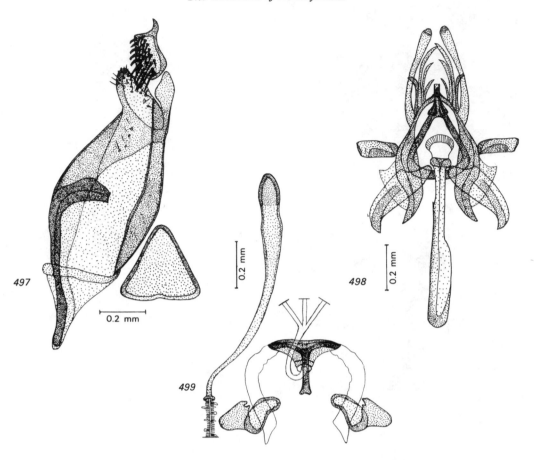

Figs. 497–499: *Pipunculopsis* n. sp.
497. gonopod; 498. aedeagus; 499. spermatheca

Desmatoneura argentifrons Williston, 1895

(Figs. 500–502)

Epandrium rounded posteriorly, concave anteriorly. Gonocoxites narrow, with long, inner apical processes. Dististyli long, narrow, with sinuate dorsal margin, pointed, with a short, subapical process. Hypandrium absent. Aedeagus conical, lacking an aedeagal process, with slightly funnel-shaped apex.

Spermathecae quite aberrant. Capsules globular; ducts short, widening proximally, striated. At their ends there is a second rounded capsule as large as the apical capsule, with a conical basal process. Ejection apparatus short, with a few large processes and large end plates bearing processes. Furca with two bars. Tergite 8 with a long, narrow apodeme with a triangular base. Acanthophorites with 4–5 long spines.

A similar duplication of the sperm capsules has been found only in the Palaearctic *Bombylius punctatus*.

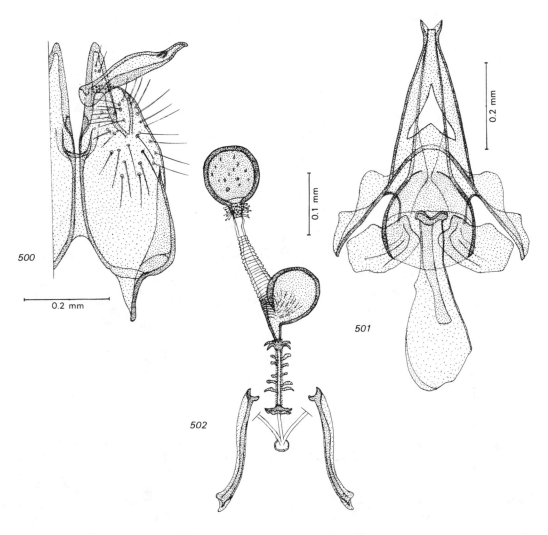

Figs. 500–502: *Desmatoneura argentifrons*
500. gonopod; 501. aedeagus; 502. spermatheca

Eucessia rubens Coquillett, 1886
(Figs. 503–505)

Epandrium rectangular, with projecting, rounded posterior corners and an indentation in the middle of the posterior margin. Cerci with numerous hairs. Gonocoxites very narrow apically, wide basally. Dististyli triangular. Hypandrium triangular, with rounded apex. Aedeagus narrow, conical, slightly curved. Aedeagal process with large, curved, pointed apical part and with two narrow, pointed processes at its base.
Spermathecae with large, oblong capsules with rounded apex. Ducts narrow. Ejection apparatus short, with short processes and large end plates bearing blunt processes. Furca

177

U-shaped, with broad basal ends with a small, inner process. Tergite 8 nearly triangular, with a very large, truncate-triangular apodeme. Acanthophorites with six very large, curved spines.

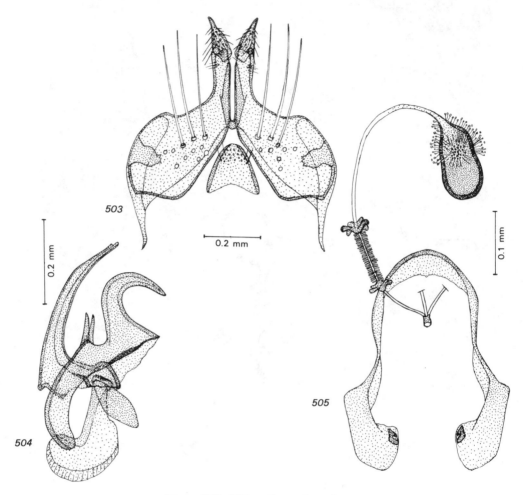

Figs. 503–505: *Eucessia rubens*
503. gonopods; 504. aedeagus, lateral; 505. spermatheca

Lomatia Meigen, 1822

L. abbreviata, conspicabilis, infernalis, lepida, and several undetermined species (Figs. 506–509).

Epandrium rounded posteriorly, with an indentation in the middle of the posterior margin and long, proximally directed basal processes. There are specific differences in form. Gonocoxites long, narrow, nearly parallel-sided, with pointed apex. Dististyli oblong, nearly parallel-sided, with an apical indentation and a short point. Hypandrium absent. Aedeagus conical, narrow, pointed. Aedeagal process absent. There is a rhomboidal or

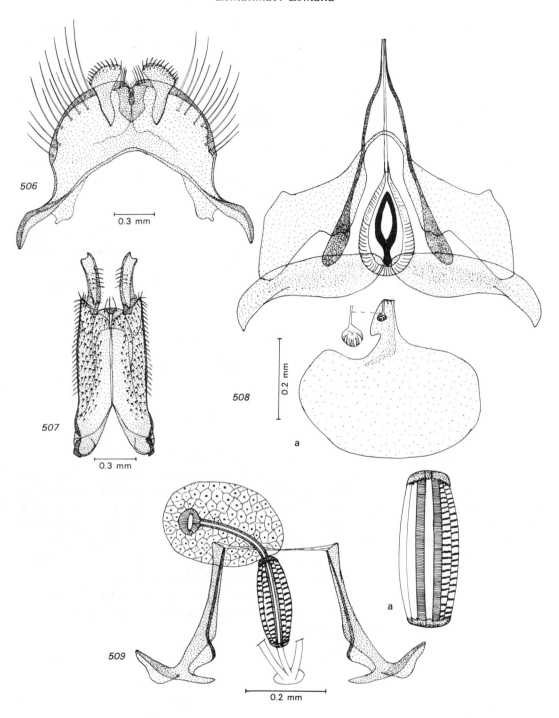

Figs. 506–509: *Lomatia infernalis*
506. epandrium; 507. gonopods; 508. aedeagus; (a) apodeme and sensilla;
509. spermatheca; (a) ejection apparatus, enlarged (fresh material)

nearly triangular, black sclerotization in the basal part of the aedeagus. This consists in fact of two processes connected with the head of the apodeme, which at first diverge and become approximated apically. This structure is probably homologous to the 'endoaedeagus' described below for *Anthrax* and *Spongostylum*. The sensilla of the apodeme are not situated at the apex of the apodeme as in most other Bombyliidae, but in a rounded structure some distance from the head (Fig. 508).

Spermathecae very short; capsules rounded, thick-walled, sometimes flattened. Ejection apparatus short, striated, without processes and with small end plates. Furca with two bars with transversely widened basal ends. Tergite 8 with a long, narrow apodeme with a wide, triangular base. Acanthophorites with 6–10 long spines with curved apex.

Anisotamia ruficornis Macquart, 1840

This species is closely related to *Lomatia* and their genitalia are very similar. There are only minor differences. The gonocoxites are truncate-triangular and the dististyli are nearly triangular, wider in the middle and with a pointed apex. Aedeagus as in *Lomatia*. Spermathecae as in *Lomatia*.

Bryodemina valida (Wiedemann, 1830), Ogcodocera analis Williston, 1901 (Nearctic); Lyophlaeba infumata Philipi, 1865, L. lugubris Rondani, 1864 (Neotropical); Comptosia apicalis Macquart, 1848, C. lateralis (Newman, 1841), Oncodosia tendens (Walker, 1849) (Australian) (Figs. 510–517)

The above genera have male genitalia closely resembling those of *Lomatia*, including the internal sclerotization of the aedeagus described for it. However, the spermathecae are different. Those of *B. valida* have very large, globular capsules with a conical base. Ducts very short. Ejection apparatus also very short, striated, without processes. End plates cup-shaped, without processes. Furca with two bars with long, lateral basal processes. Acanthophorites with a dense row of about 20 long spines. The spermathecae of *O. analis* have cylindrical capsules with a rounded, slightly widened apex. Ducts, ejection apparatus and spines on the acanthophorites as in *Bryodemina*. The spermathecae of *Lyophlaeba, Comptosia* and *Oncodosia* are similar but the capsules are of varying width and length. Tergite 8 with a long, narrow apodeme with a triangular base. Acanthophorites with 5–8 large, curved spines. The black sclerotization described for *Lomatia*, which is located in the base of the aedeagus and connected with the head of the apodeme, is also present, but the apices of the processes are not approximated.

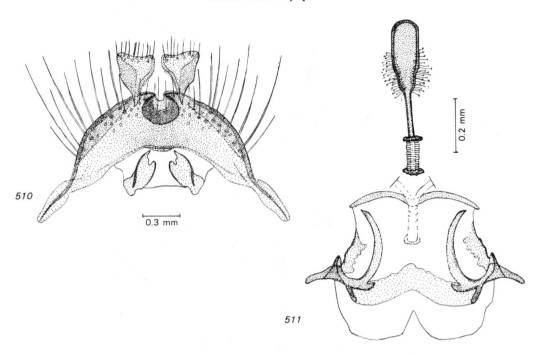

Figs. 510–511: *Lyophlaeba infumata*
510. epandrium; 511. spermatheca

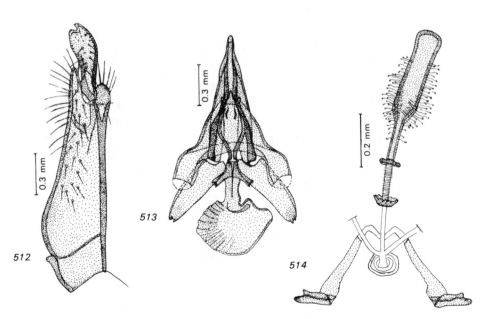

Figs. 512–514: *Lyophlaeba lugubris*
512. gonopod; 513. aedeagus; 514. spermatheca

181

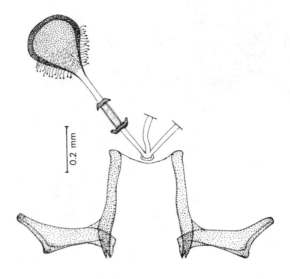

Fig. 515:   *Bryodemina valida*, spermatheca

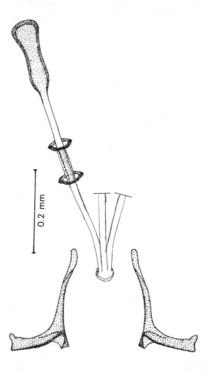

Fig. 516:   *Ogcodocera analis*, spermatheca

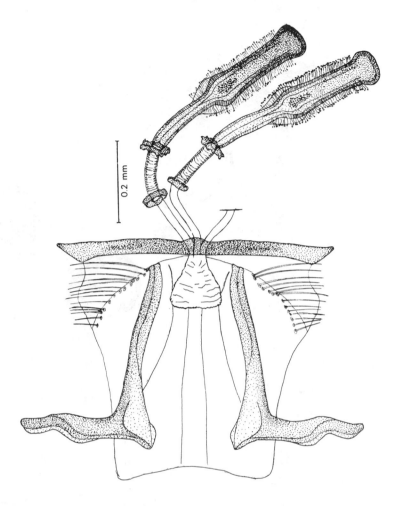

Fig. 517 : *Comptosia apicalis*, spermathecae

Chionamoeba Sack, 1909

C. choreutes Bowden, 1964, *nivea* (Rossi, 1970) (Figs. 518, 519)

The systematic position of the genus is doubtful and its composition is apparently heterogenous. Hull (1973) placed it in the Lomatiinae, but the wing venation and the antennae with hairs at the apex of segment 3 (at least in some species) resemble the Anthracinae more closely. Only species closely resembling the type species *nivea* in the male genitalia, the bulging frons and the narrow ventral part of the face are considered here as belonging to this genus. Some species (*sabulonis*, (?) *lepida*) which have a different frons and different genitalia have been placed in the genus. The status of *semirufa* is also doubtful although the head resembles that of *nivea*. These species may have to be placed in a different genus or genera. Bezzi (1924) considered *semirufa* as an *Anthrax*.

183

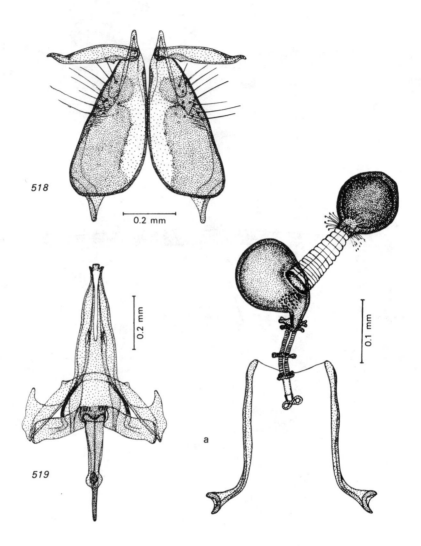

Figs. 518–519: *Chionamoeba nivea*
518. gonopods; 519. aedeagus; (a) *Chionamoeba choreutes*, spermatheca

Epandrium rounded posteriorly; cerci large. Gonocoxites triangular, with pointed apex. Hypandrium absent. Dististyli long, narrow, with sinuate dorsal side and slightly curved, pointed apex. Aedeagus conical; sheath extending to apex of aedeagus. Aedeagal process absent.

The spermathecae of *C. choreutes* consist of a large, globular capsule and a short, wide, striated duct which ends in a second, even larger capsule. Ejection apparatus very short, striated, with large end plates with blunt processes. The ducts end in sclerotized rings in the vagina.

Chionamoeba semirufa Sack, 1909
(Figs. 520, 521)

The head and antennae resemble those of *nivea* but the male genitalia are different. Epandrium rounded posteriorly. Gonocoxites very broad, triangular. Dististyli short, wide, their base with spines, with pointed apex and a short apical process. Hypandrium small, broadly triangular. Aedeagus conical; aedeagal process V-shaped, long, curved, with bifid apex, its processes pointed. Aedeagus with internal denticles. Female not examined.

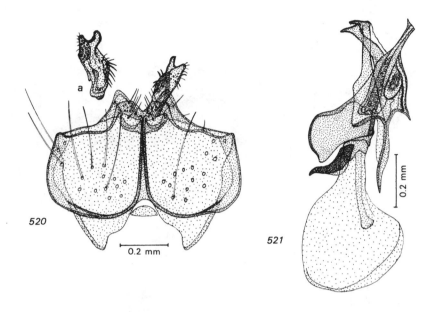

Figs. 520–521: *Chionamoeba semirufa*
520. gonopods; (a) dististylus, lateral; 521. aedeagus, lateral

Chiasmella Bezzi, 1924

C. sabulonis (Becker, 1906), *C.* (?) *lepida* (Hermann, 1907), and a new species (Figs. 4b, 522–532)

These species differ from *nivea* in the form of the head (frons not bulging, face less narrow ventrally) and in the genitalia.

Epandrium nearly trapezoidal, with concave basal margin. Gonocoxites very broad, truncate-triangular. Dististyli triangular, with broad base with a large, vertical, rounded plate in *sabulonis*, the plate smaller or rectangular in the other species. Hypandrium absent. Aedeagus conical, with more or less wide apex. Basal plates very large, rounded. Aedeagal process of varying form, with broad, recurved apex with processes in (?) *lepida*, with a single pointed denticle in *sabulonis* and the new species. There is an 'endoaedeagus' with three ridges as in the Anthracinae.

185

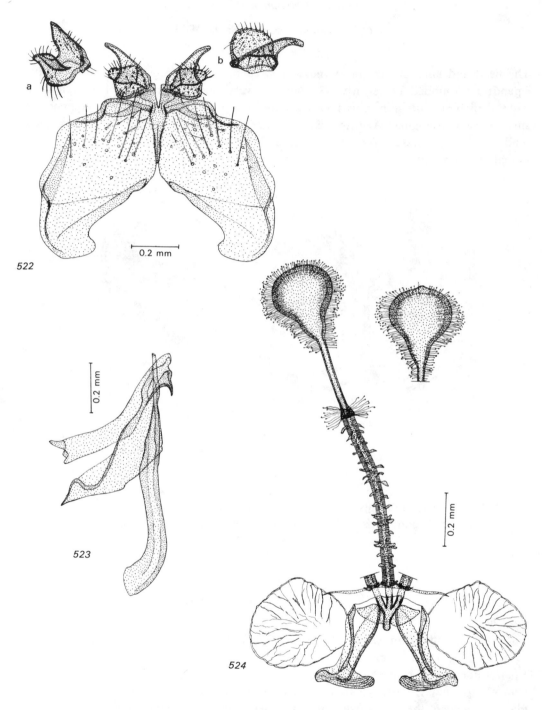

Figs. 522–532: Species apparently not belonging to *Chionamoeba*

Figs. 522–524: *Chiasmella sabulonis*

522. gonopods; (a) and (b) different aspects of dististylus; 523. aedeagus, lateral; 524. spermatheca

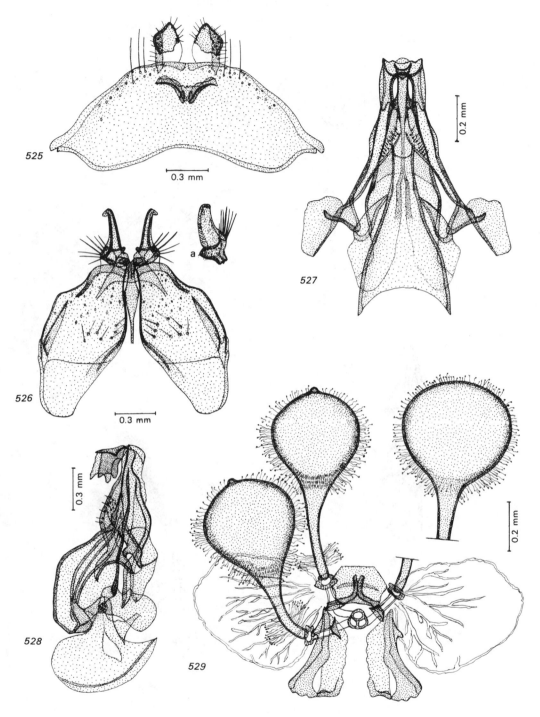

Figs. 525–529: (?) *Chiasmella lepida*
525. epandrium; 526. gonopods; (a) dististylus, lateral; 527. aedeagus;
528. same, lateral; 529. spermathecae

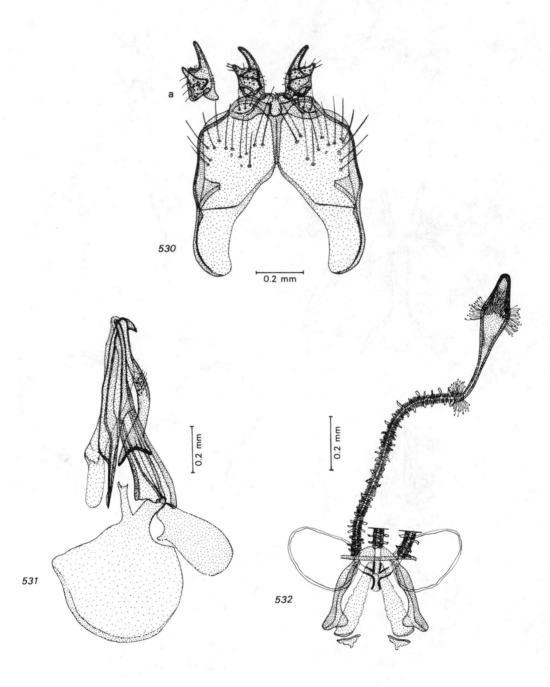

Figs. 530–532: (?) *Chiasmella* sp. no. 1
530. gonopods; (a) dististylus, different aspect; 531. aedeagus; 532. spermatheca

Spermathecae of varying form and size. The capsules of (?) *lepida* are very large, globular, with a conical base. Ducts short, sclerotized; ejection apparatus very short, without processes and with cup-shaped end plates. Furca with two broad bars and a sclerite with a T-shaped sclerotization between their apical ends. The capsules of *sabulonis* are similar, but smaller; however, the ejection apparatus is much longer, with short and longer processes and small end plates. The sperm capsules of the new species are rhomboidal; ducts short; ejection apparatus long, narrow, with short and longer processes, lacking distinct end plates but with a conical apical sclerotization. There are two large, crinkled, membranous structures lateral to the furca, the function of which is not clear. Acanthophorites with 2–4 long, thin spines in the three species. Tergite 8 with a short, narrow apodeme.

All the above characters, including the presence of an 'endoaedeagus', suggest that the species are closely related to the Anthracinae.

New Genus near Chionamoeba
(Figs. 533–537)

The specimens were bred from pupae of Myrmeleonidae. They resemble *Chionamoeba* in the slightly bulging frons and the ventrally narrow face, but differ in the form of segment 3 of the antennae, which has a simple style, without hairs at the apex, and in the distinctly different genitalia of both sexes.

Epandrium short, trapezoidal, with short, lateral basal processes. Gonocoxites fused. Cerci with a pigmented area. Hypandrium apparently fused with the gonocoxites. Dististyli short, broad, with a vertical rounded plate at the base resembling that in the species of *Chiasmella*, and a short apical process. Aedeagus conical; sheath with rounded lateral shoulders. Aedeagal process very wide, flattened, with rounded apex, without differentiations. Basal plates short and broadly rounded.

Spermathecae with oblong, rounded capsules. Ducts very short, wide, membranous, with a long, asymmetrical widening. From this extends a short duct, with a sclerotization at the beginning, to the ejection apparatus which is short, striated and has small end plates. The basal end plate is rounded, cup-shaped and extends in a thin duct to the vagina. Furca rectangularly U-shaped, its apex with a median ridge. Tergite 8 with a short, narrow apodeme. There are no acanthophorites, and distinct setae are absent on tergite 9 or there are one or two long, thin setae on one or both sides.

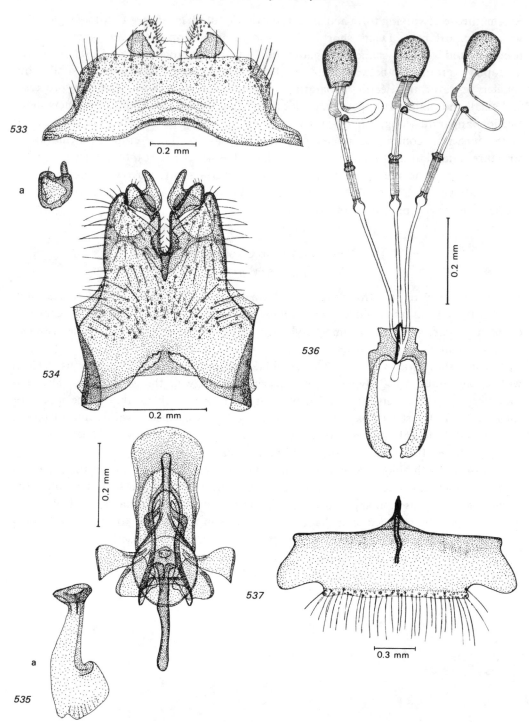

Figs. 533–537: New genus (near *Chionamoeba*)
533. epandrium; 534. gonopods; (a) dististylus, different aspect;
535. aedeagus, dorsal; (a) apodeme; 536. spermathecae; 537. tergite 8 of female

Antonia Loew, 1856

A. fedchenkoi, suavissima, xanthogramma, and an undescribed species (Figs. 538–543)

The genus *Antonia* is usually placed in the Lomatiinae. However, it is quite aberrant and cannot be placed in any of the accepted subfamilies.

The head is typical for the Tomophthalmae, with a deeply invaginated occiput, two occipital foramina, an indentation of the hind margin of the eyes and a bisection line. However, the abdomen of the female is quite undifferentiated and there is only a single spermatheca.

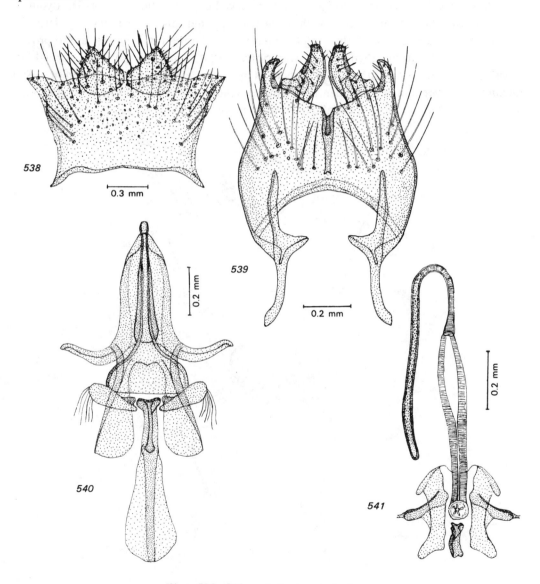

Figs. 538–541: *Antonia suavissima*
538. epandrium; 539. gonopods; 540. aedeagus; 541. spermatheca

Epandrium nearly rectangular, slightly wider posteriorly, with pointed or rounded posterior corners. Gonocoxites broadly truncate-triangular, fused, but with a median suture, a long, narrow basal process and a lateral, slightly curved, pointed apical process. Dististyli triangular, slightly curved, with a rounded apex in *suavissima*, long and narrow, with a wider, recurved apex in *fedchenkoi*. Hypandrium absent. Aedeagus with long, narrow, conical apical part and wide basal part. Aedeagal process wide, flattened, with a broadly rounded apex in *fedchenkoi*, slightly tapering in *suavissima*. Apodeme with two lateral transverse plates, so that its transverse section is cruciform.

The single spermatheca has a long, tubular, recurved capsule which passes directly into a wide, striated, membranous duct with a plicated opening into the vagina. A sclerotized ejection apparatus is apparently absent. Furca with two broad bars with a lateral process and a median sclerite of irregular form posteriorly. Tergite 8 not differentiated, normal, not invaginated. Tergite 9 with two short processes in *suavissima*, with two long, thin processes in *xanthogramma* and in an unidentified female. Acanthophorites and spines absent. Hesse (1956) illustrated the female genitalia of *A. xanthogramma*.

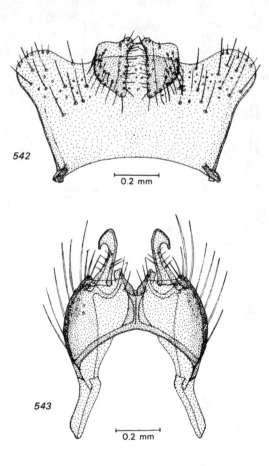

Figs. 542–543: *Antonia fedchenkoi*
542. epandrium; 543. gonopods

ANTHRACINAE Latreille, 1804

Anthrax Scopoli, 1763

A. aethiops, anthrax, binotatus, candidapex, comatus, fuscipennis, niger, sticticus, trifasciatus, varius, and several undescribed species (Palaearctic); *A. aterrimus, cascadensis, daphne, limatulus, simson habrosus* (Nearctic); *A. maculatus* (Australian) (Figs. 4a, 544–590)

There have been numerous attempts to divide this large genus into genera, subgenera and species groups. Most of these names have been abandoned, except the division into *Anthrax* and *Spongostylum*. The characters mainly used for the distinction of these two genera were the structure of the antennae and the venation and coloration of the wings. Bezzi (1924) and Engel (1936) stated that segment 2 of the antennae of *Spongostylum* is cup-shaped and segment 3 is fitted into it, while in *Anthrax* segment 2 is rounded and segment 3 is not fitted into it. However, there are species of *Anthrax* with a cup-shaped segment 2 (*A. aethiops* and related species). Wings with a more or less extensive black pattern have been found so far only in *Anthrax*, but there are also species with completely or almost completely clear wings; some species of *Spongostylum* have dark spots on the wings, while others have clear wings. There are species with two and three submarginal cells in both genera.

The genitalia have not been used in the past to distinguish the two genera. Only recently, Marston (1970) gave illustrations of the male and female genitalia of American species of *Anthrax* and Mühlenberg (1970) described the spermathecae of *A. anthrax*.

The structure of the genitalia permits the distinction of two groups in the Palaearctic species of the genus.

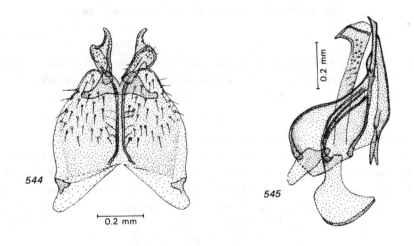

Figs. 544–545: *Anthrax anthrax*
544. gonopods; 545. aedeagus

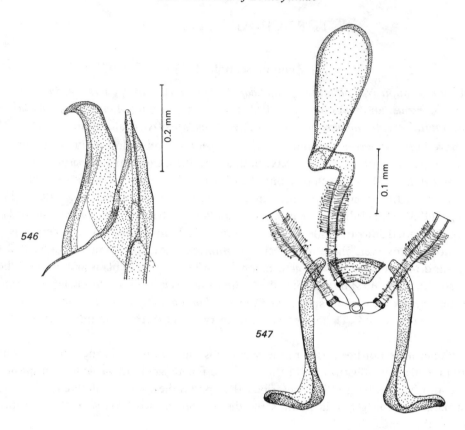

Figs. 546–547: *Anthrax aethiops*
546. aedeagus, apical part; 547. spermatheca

1. *A. anthrax* group. Cerci without a distinct armature of spines. Gonocoxites truncate-triangular. Dististyli triangular. Aedeagal process simple, with hooked apex. Spermathecae with long, club-shaped, tubular capsules. Ejection apparatus very short, weakly sclerotized.

2. *varius* group. Cerci with a distinct armature of short, black spines. Gonocoxites with broad base and much narrower apical part with a more or less long apical process. Dististyli of markedly varying form, with distinct specific differences. Aedeagal process with complicated apical differentiations. Spermathecae with strongly sclerotized capsules of specific form and a longer, more strongly sclerotized ejection apparatus. This group closely resembles the Nearctic *albofasciatus* group in the form of the gonocoxites and the spermathecae. Apparently, there are additional groups in America.

The position is particularly complicated in the *fuscipennis* group. There are apparently several species with a similar wing pattern. One form has a black plumula, another a white plumula, but different genitalia were found in specimens with the same plumula in both

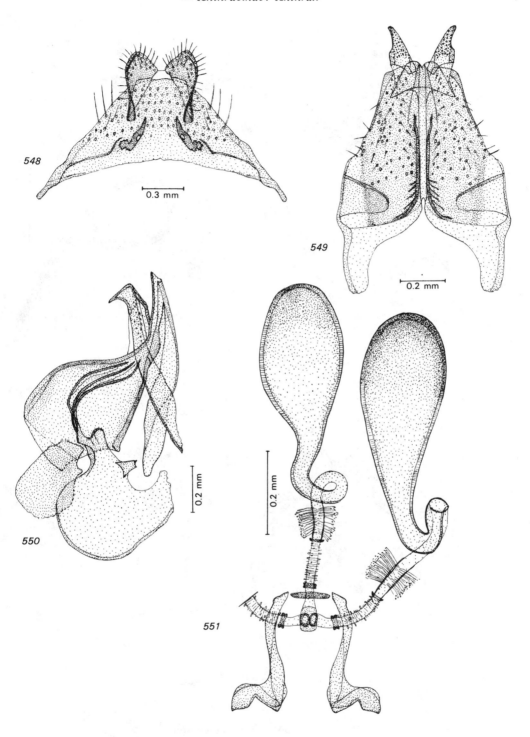

Figs. 548–551: *Anthrax* sp. no. 1
548. epandrium ; 549. gonogpods ; 550. aedeagus ; 551. spermathecae

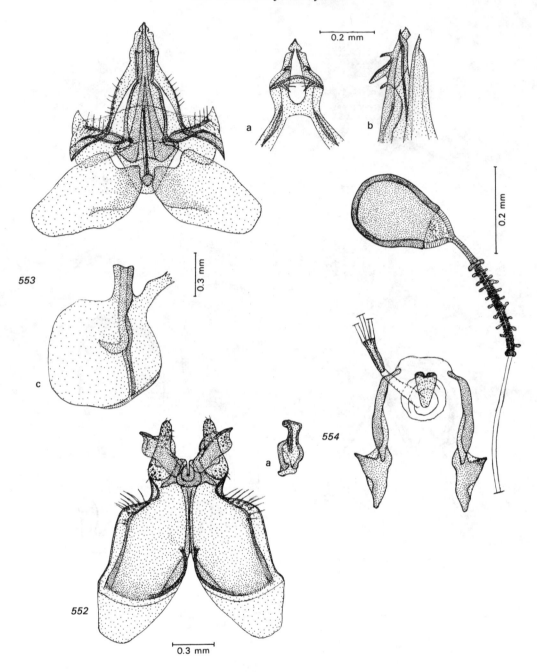

Figs. 552–554: *Anthrax candidapex*
552. gonopods; (a) dististylus, different aspect
553. aedeagus; (a) same, apex enlarged; (b) same, lateral; (c) apodeme; 554. spermatheca

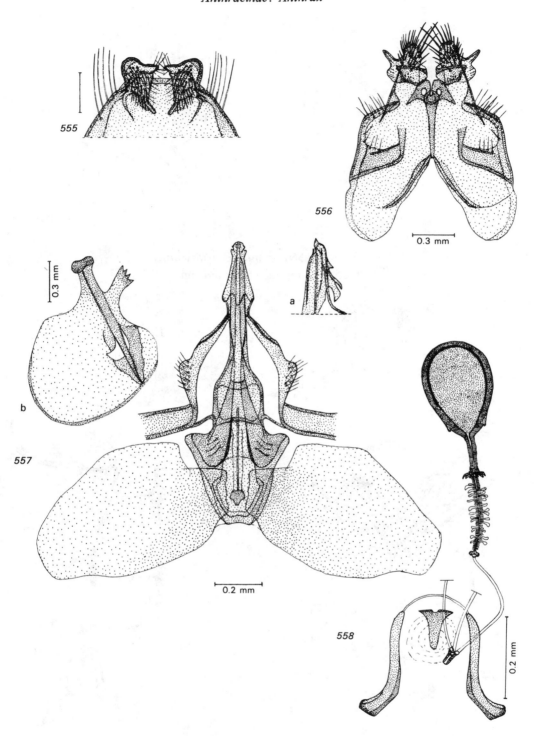

Figs. 555–558: *Anthrax niger*

555. cerci; 556. gonopods; 557. aedeagus; (a) apex, lateral; (b) apodeme; 558. spermatheca

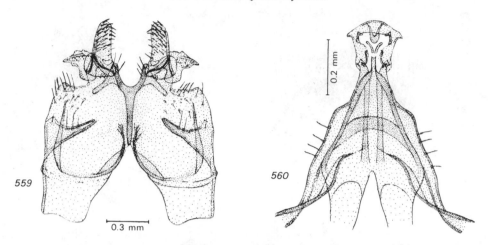

Figs. 559–560: *Anthrax trifasciatus*
559. gonopods; 560. aedeagus

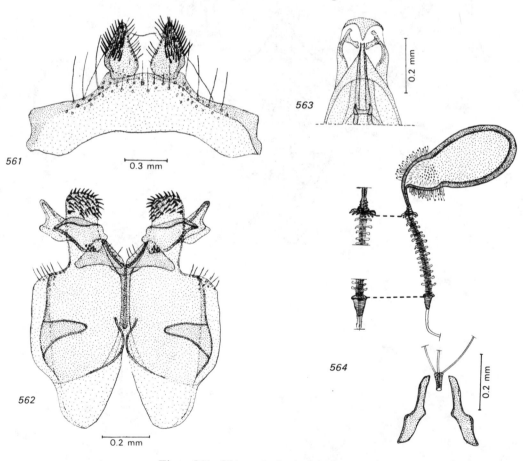

Figs. 561–564: *Anthrax binotatus*
561. epandrium; 562. gonopods; 563. aedeagus, apical part; 564. spermatheca

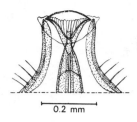

Fig. 565: *Anthrax* sp. no. 2 (near *binotatus*), aedeagus, apex

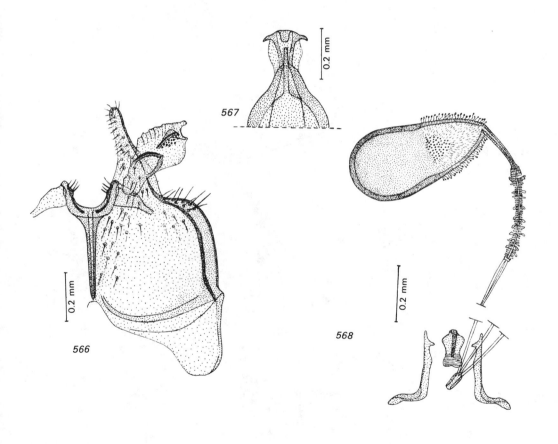

Figs. 566–568: *Anthrax fuscipennis*, type 1 (black plumula)
566. gonopods; 567. aedeagus, apex; 568. spermatheca

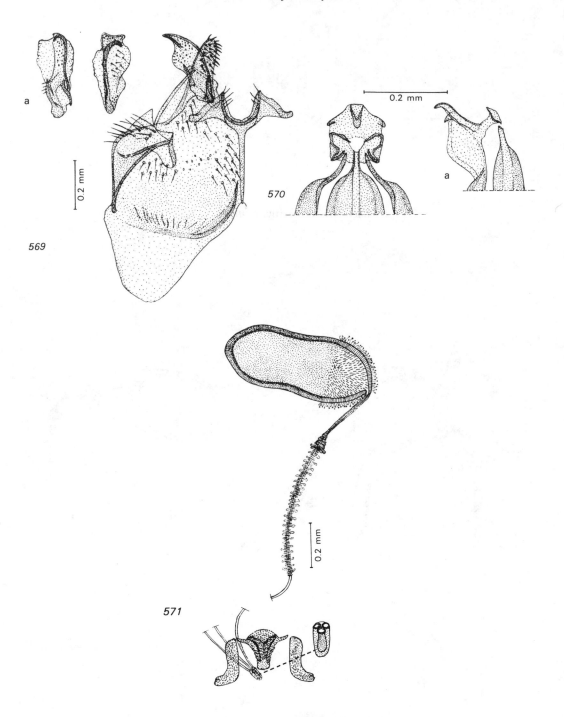

Figs. 569–571: *Anthrax fuscipennis*, type 2 (white plumula)
569. gonopod; (a) dististylus, different aspects;
570. aedeagus, apex; (a) same, lateral; 571. spermatheca

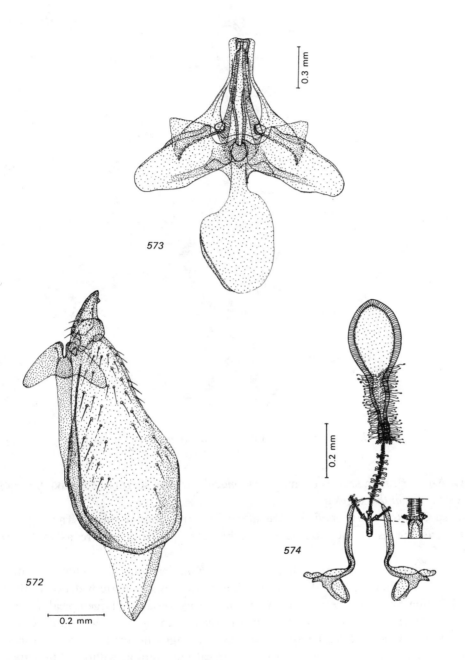

Figs. 572–574: *Anthrax comatus*
572. gonopod; 573. aedeagus; 574. spermatheca

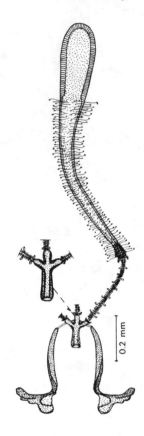

Fig. 575:   *Anthrax* sp. no. 3 (near *comatus*), spermatheca

forms. An Australian species examined (*A. maculatus*) has spermathecae like those of the species of the *anthrax* group.

Some species (*comatus, sticticus*) are apparently transitional forms. The gonocoxites of *sticticus* are as in the *anthrax* group, but the spermathecae are as in the *varius* group.

An undescribed species resembling *A. chionanthrax* in some characters has different spermathecae. They are very short, with tubular capsules which are widened in the basal part. Ejection apparatus very short, with small processes and a large apical and a small basal end plate. The part of the ducts before their opening in the vagina is sclerotized. Furca with two long, lateral basal processes. Aedeagus normal; aedeagal process not reaching apex of aedeagus, broad, with an apical indentation, without differentiations. Gonocoxites truncate-triangular; dististyli triangular. Hypandrium very small.

The American species have been revised by Marston (1963, 1970). *A. simson habrosus* (*tigrinus* group) differs distinctly from the other species in several characters. Dististyli oblong-triangular, with a curved dorsal process. Aedeagus very long, thin, S-curved, its basal part recurved posteriorly. Sheath very wide, membranous, crinkled. Aedeagal process very long, with oblong-triangular apex. It resembles the Palaearctic *Spongostylum*

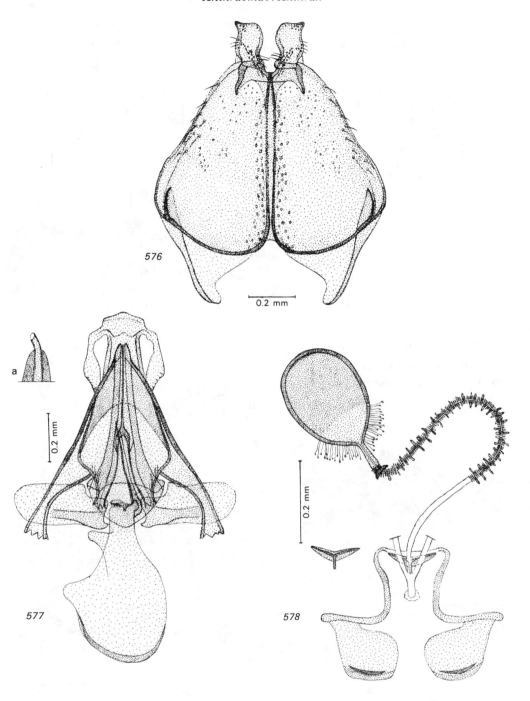

Figs. 576–578: *Anthrax sticticus*
576. gonopods; 577. aedeagus; (a) apex of aedeagus; 578. spermatheca

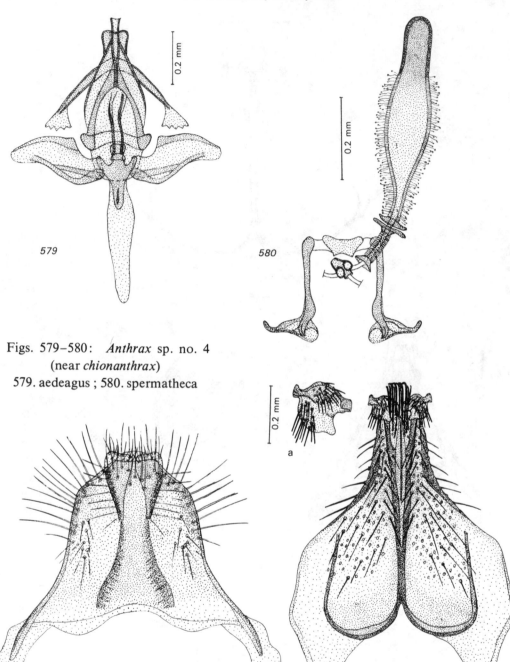

Figs. 579–580: *Anthrax* sp. no. 4
(near *chionanthrax*)
579. aedeagus ; 580. spermatheca

Figs. 581–582: *Anthrax aterrimus*
581. epandrium; 582. gonopods; (a) dististylus, enlarged

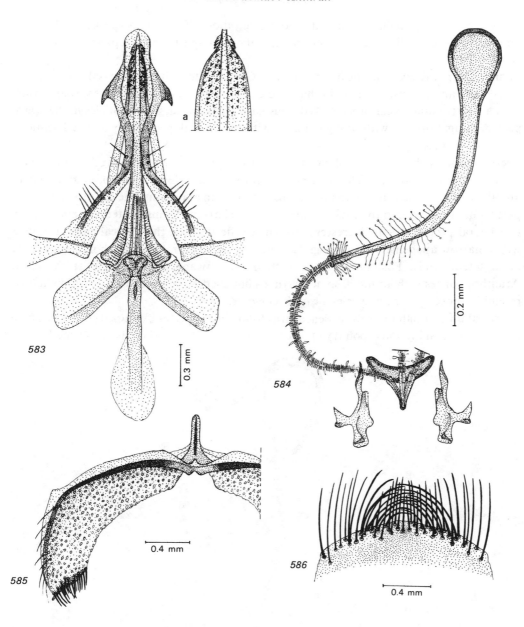

Figs. 583–586: *Anthrax aterrimus*
583. aedeagus; (a) apex of aedeagus, enlarged; 584. spermatheca;
585. tergite 8 of female; 586. posterior sternite of female, apex

etruscum in some characters, but the spermathecae are different. The drawing of the aedeagus given by Marston (1970) depicts only the aedeagal process; the aedeagus itself is not shown. A similar aedeagus has been illustrated by Hesse (1956, Figs. 166, 167), for the Ethiopian species *Argyramoeba punctipennis* and *punicisetosa*.

A. aterrimus has genitalia differing distinctly from those of the other species. Hull (1973) unfortunately gives illustrations only of this aberrant species as a representative of the genus.

Epandrium relatively long, nearly triangular. Gonocoxites with wide basal part, narrow apical part and a brush of hairs in the middle at the apex. Dististyli very short and wide, with a short, blunt apical process. Aedeagus narrowly conical, with denticles in the apical part. Aedeagal process with triangular apex with two lateral, proximally directed denticles on its wide, basal part.

Spermathecae with nearly globular capsules which continue in a narrowing sclerotized part to the ejection apparatus. This is narrow, with short and longer processes and a small apical end plate. Ducts to the vagina absent; the ejection apparatus extends to a sclerotized part before opening into the vagina. Furca with two bars, widening posteriorly and bearing two lateral processes with a crescent-shaped sclerite between their apical ends. Tergite 8 with a narrow apodeme with a triangular base and with several rows of short spines at the basal lateral ends. Posterior sternite with dense, inwards curved setae at the apex. Acanthophorites with about 20 spines having a flattened, widened apex. The spermathecae resemble those of a new species near *A. comatus*.

A special differentiation of the aedeagus was found in all species examined (Fig. 4a). There is a closed, membranous, pointed tube extending from the head of the apodeme both in

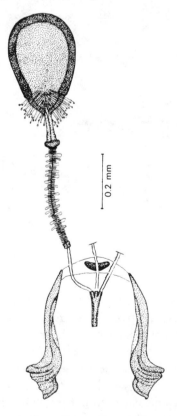

0.2 mm

Fig. 587: *Anthrax limatulus*, spermatheca

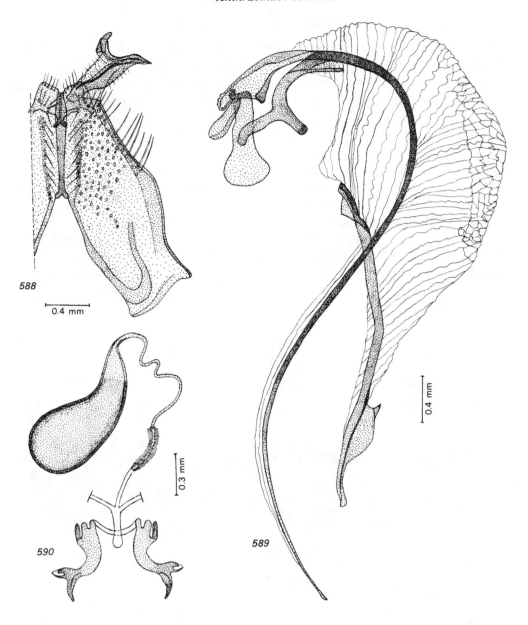

Figs. 588–590: *Anthrax simson habrosus*
588. gonopod; 589. aedeagus; 590. spermatheca

Anthrax, Spongostylum and the species of *Chiasmella*. The tube has 2–3 narrow, sclerotized ridges in most species (apparently none in *A. simson habrosus*). This is apparently homologous to the 'endoaedeagus' described for the Asilidae (Theodor, 1976). The two black processes which form a rhomboidal figure in *Lomatia* and related genera are apparently a similar structure.

207

Spongostylum* Macquart, 1840

S. bilineatum, candidum, etruscum, hippolyta, isis, mixtum, ocyale, perpusillum, tripunctatum, and several undescribed species (Figs. 591–614)

This genus is closely related to *Anthrax* and has been considered as a subgenus of *Anthrax* by various authors. The characters used in the past do not permit separation from *Anthrax* as discussed above, but the structure of the aedeagus and particularly the spermathecae differ so markedly from those of *Anthrax* that it seems justified to maintain the genus.

There is confusion about the type species of the genus. Macquart mistakenly considered the type species *mystaceum* as South American. However, Séguy (1938) stated that this species was in fact collected in Sinai; it is most probably *tripunctatum* (not *punctipenne* as suggested by Séguy, which is a South African species).

The species are all very similar, but the genitalia show distinct specific characters. Thus, specimens identified as *S. isis* by external characters showed three different types of male genitalia.

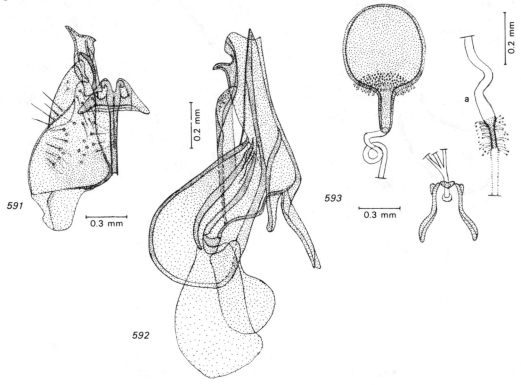

Figs. 591–593: *Spongostylum tripunctatum*
591. gonopod; 592. aedeagus; 593. spermatheca; (a) sclerotization of duct

---

* The emendation of *Spogostylum* to *Spongostylum* is correct. The transliteration of the double gamma into Latin is 'ng'. See *Code of Zool. Nomenclature* (1964), p. 133, footnote.

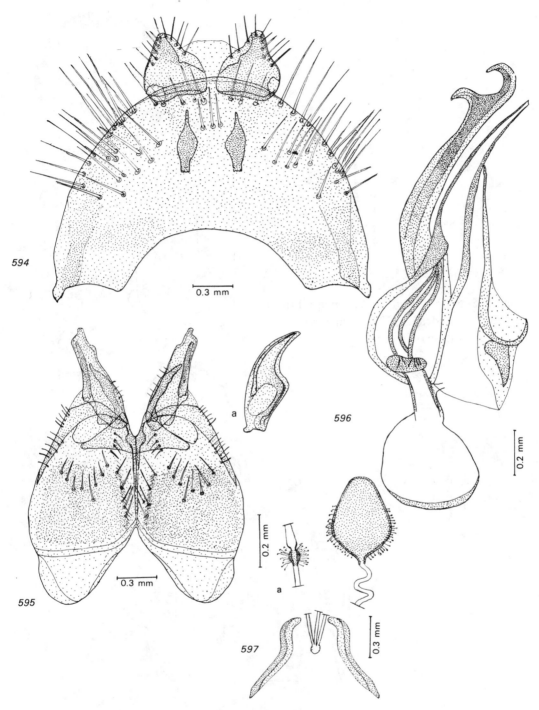

Figs. 594–597: *Spongostylum ocyale*
594. epandrium; 595. gonopods; (a) dististylus, lateral; 596. aedeagus;
597. spermatheca; (a) sclerotization of duct

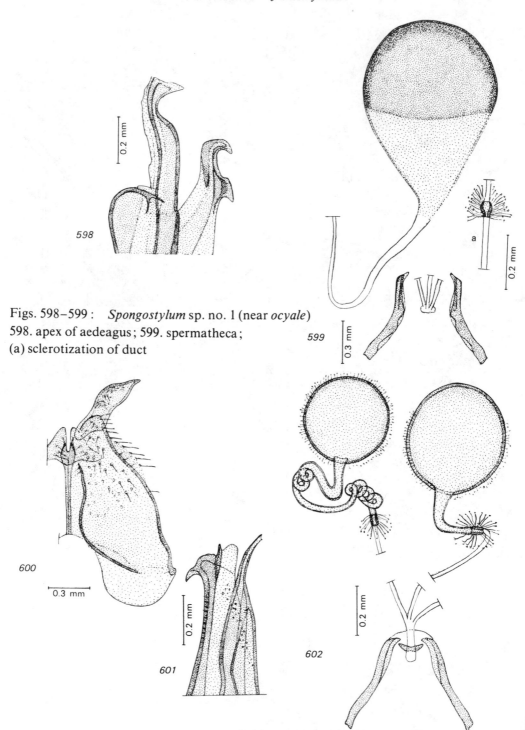

Figs. 598–599 : *Spongostylum* sp. no. 1 (near *ocyale*)
598. apex of aedeagus; 599. spermatheca;
(a) sclerotization of duct

Figs. 600–602: *Spongostylum candidum*
600. gonopod; 601. apex of aedeagus; 602. spermatheca

210

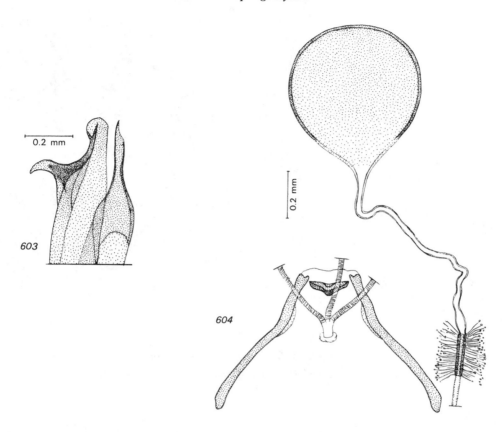

Figs. 603–604: *Spongostylum hippolyta*
603. apex of aedeagus; 604. spermatheca

Epandrium usually short, trapezoidal, rounded posteriorly, with blunt posterior corners. Gonocoxites truncate-triangular; dististyli oblong-triangular, with slightly rounded sides but of distinctive form in some species, nearly parallel-sided in the greater basal part in *mixtum*. Aedeagus conical. Aedeagal process of distinctive form in most species, with two apical hooks in *ocyale, bilineatum* and others, with one hook in *isis*; the hooks and apex may be of different form and size. The aedeagal process of *ocyale* illustrated by Engel (1936) shows only a single hook (origin of specimen not given). There are apparently several similar species which have been identified as *ocyale*, and its 'great variation' was mentioned by Austen (1937) who also mentioned specimens from Iraq 'with almost exclusively black hairs on the abdomen'. A species from Israel identified as *ocyale* had distinctly different genitalia. An 'endoaedeagus' as described for *Anthrax* is present in all species examined, with 2–3 ridges in most species, but without such ridges in *etruscum*.

The spermathecae differ distinctly from those of all species of *Anthrax* examined. The capsules are very large, either of irregular rhomboidal form, with very thin walls (*ocyale*), or rounded and more strongly sclerotized (*isis, hippolyta, tripunctatum*). They may be of equal

211

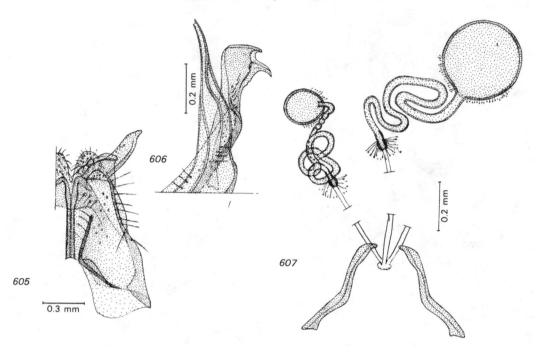

Figs. 605–607 : *Spongostylum mixtum*
605. gonopod; 606. apex of aedeagus; 607. spermathecae

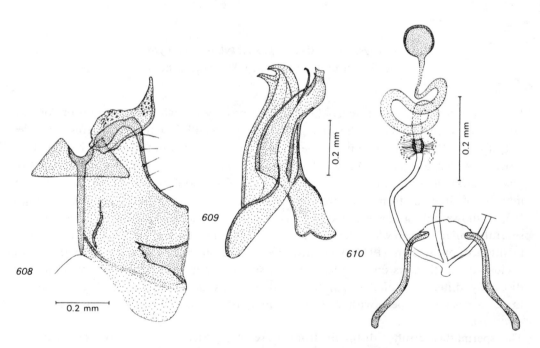

Figs. 608–610: *Spongostylum perpusillum*
608. gonopod; 609. aedeagus; 610. spermatheca

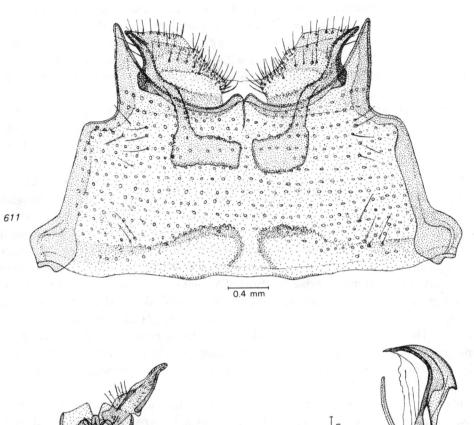

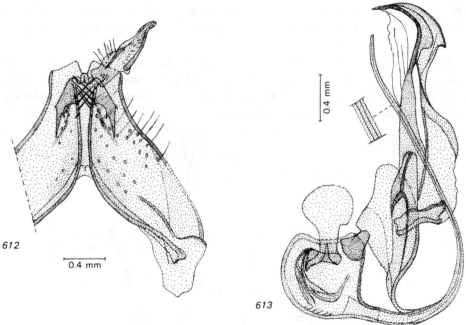

Figs. 611–613: *Spongostylum (Satyramoeba) etruscum*
611. epandrium; 612. gonopod; 613. aedeagus

size or the median spermatheca is much larger than the lateral ones (*mixtum*). The ducts are convoluted in some species (*mixtum, perpusillum*). They end in a small sclerotized part on which the gland is situated. This part may be long and cylindrical or short and ovoid. An ejection apparatus is absent. The ducts are without differentiations. Furca with two thin bars with specific differences. Tergite 8 with a short, narrow apodeme. Acanthophorites with 5–7 long spines, 3–4 in *perpusillum*.

S. (Satyramoeba) etruscum Fabr. differs distinctly from the other species in external characters (black coloration, large size, etc.) and in the very characteristic genitalia. *Satyramoeba* was considered as a subgenus of *Spongostylum* by Engel, but it may belong to a different genus.

Epandrium rectangular, with long, pointed, posterior lateral processes. Cerci of characteristic form. Gonocoxites relatively narrow; dististyli irregularly triangular. Aedeagus with large, rounded basal part; apical part very long, thin, slightly flattened; apodeme small, directed posteriorly, in the opposite direction to the narrow part. Aedeagal process long, V-shaped, with long, curved, pointed apical process. The aedeagus resembles that of *A. punctipennis* illustrated by Hesse (1956, p. 453) and that of *A. simson habrosus*. The spermathecae resemble in principle those of *Spongostylum*. The capsules are very large (1.5 × 1 mm), thin-walled, forming oval, membranous sacs, with a conical sclerotization at the base. Ducts short, with numerous canaliculi of the gland; sclerotized part short, cylindrical. There is a second small sclerotized part further proximally. Ejection apparatus absent. Furca with two bars, curved and widened posteriorly. Tergite 8 with a short, narrow apodeme. Acanthophorites with 28–30 long, black spines with widened, curved, clear apex.

As stated above, the aedeagus of *etruscum* resembles that of *A. simson habrosus*, but the spermathecae of the latter species are different. They are pear-shaped, sclerotized, and an ejection apparatus is present (Fig. 590). *S. etruscum* also resembles *A. simson habrosus* in the numerous spines on the acanthophorites.

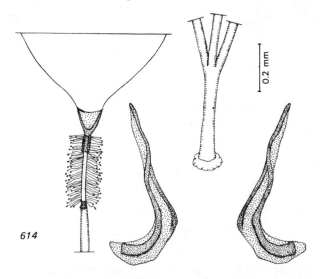

Fig. 614:   *Spongostylum (Satyramoeba) etruscum*, spermatheca

EXOPROSOPINAE Becker, 1912

Exoprosopa Macquart, 1840

About 30 Palaearctic species, including the type species *capucina* and several undescribed species; *E. caliptera, divisa, doris, rostrifera* (Nearctic); *E. fenestrata, latelimbata* (Australian) (Figs. 615–648)

This large genus contains very different forms, particularly in the coloration of the wings.

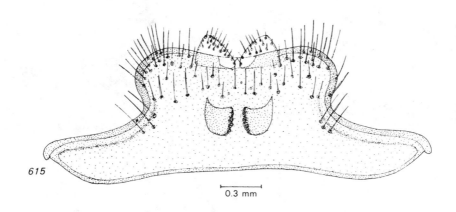

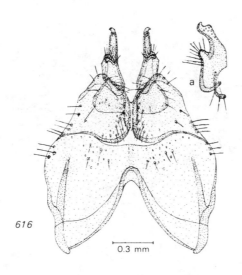

Figs. 615–616: *Exoprosopa capucina*
615. epandrium; 616. gonopods; (a) dististylus, lateral

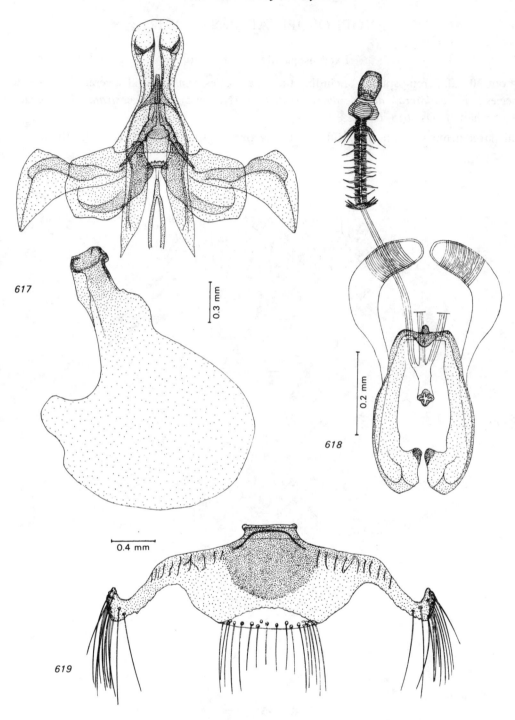

Figs. 617–619: *Exoprosopa capucina*
617. aedeagus; 618. spermatheca; 619. tergite 8 of female

There have been several attempts to divide the genus on the basis of external characters, mainly wing venation and coloration. These classifications have not been very successful and were rejected by Paramonov (1928). Many of the characters used proved highly variable. Some species groups have recently been classed as separate genera (e.g., the *stupida* group as the genus *Micromitra* Bowden).

It was hoped that the genitalia would provide characters for a more definite classification, but this has not been the case. The male genitalia are all of a similar type, with specific differentiations. The spermathecae vary distinctly in form but their differentiation is also mainly of specific rank. They are distinctly different in *Micromitra* and show group characters in the subgenus *Heteralonia*.

Epandrium either trapezoidal or rounded posteriorly, with lateral basal processes which may be short or long and curved. Epandrium of *E. minos* very short, with very long, pointed basal processes. It is strongly convex, bulging ventrally in the *tamerlan* group. The gonocoxites are fused in most species, truncate-triangular, with a markedly bulging basal part in the *tamerlan* group. Dististyli more or less triangular, with a curved, pointed apical part and a more or less long basal process in some species so that they appear bifid. Aedeagus conical, more or less long and wide, with a pointed lateral process in the middle on each side in some species (*deserticola, conspicienda, bagdadensis*). Aedeagal process flattened, with specific apical differentiations (*grandis, rivulosa*) or without such differentiations. It bears a more or less long ventral process in the *tamerlan* group.

The spermathecae show a surprising variation. There are six distinctly different types in the Palaearctic species.

Type 1. Capsules small, rounded, with a membranous, striated part at the base (*capucina* and numerous others).

Type 2. Capsules tubular, more or less long and thick, with pointed apex [*Heteralonia* according to Bowden (1975b)].

Type 3. Capsules short, tubular, with rounded apex. Membranous part at the base very large (n.sp. no. 1).

Type 4. Capsules very small, rounded, with a small membranous widening at the base. Ducts very short. Ejection apparatus very long; the short apical part lacking processes; the other part with short and longer processes, reaching the sclerotized proximal part of the ducts, with only an apical end plate. Common duct very wide, membranous, crinkled (*jaccha*).

Type 5. Capsules forming conical tubes which are sclerotized apically, membranous and curved basally and with inner differentiations. Ejection apparatus short, with small end plates (*tamerlan* group).

Type 6. Capsules broadly rounded, with an apical point. At its base is a large, rounded, membranous part with a sclerotized basal part. Ducts to ejection apparatus absent. Apical part of ejection apparatus without processes; other part with short processes. Apical end plate large, cup-shaped; basal end plate shallow, with processes (*minos*).

There are transitions and additional differentiations not mentioned in the above list. Types 1 and 2 are relatively common; the other types were found only in one or two

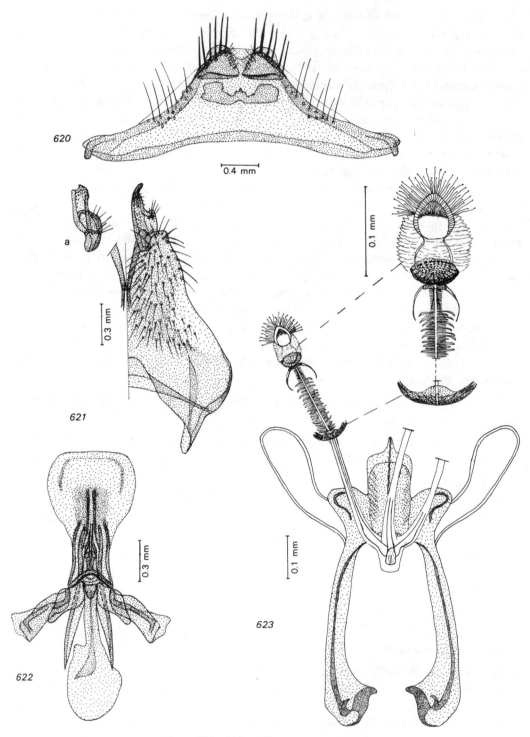

Figs. 620–623: *Exoprosopa minos*
620. epandrium; 621. gonopod; (a) dististylus, lateral; 622. aedeagus; 623. spermatheca

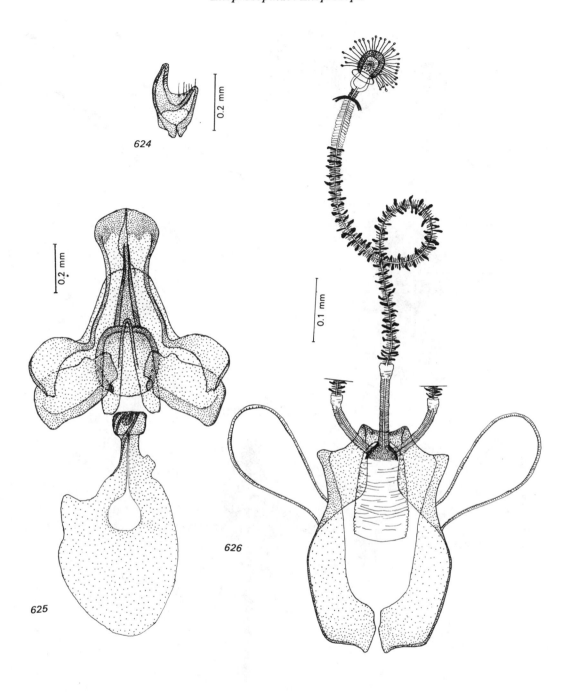

Figs. 624–626: *Exoprosopa jaccha*
624. dististylus; 625. aedeagus; 626. spermatheca

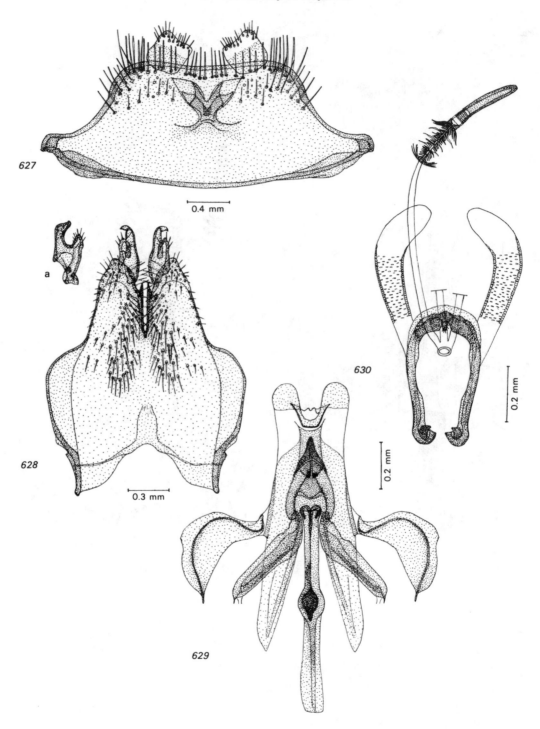

Figs. 627–630: *Exoprosopa grandis*
627. epandrium; 628. gonopods; (a) dististylus, lateral; 629. aedeagus; 630. spermatheca

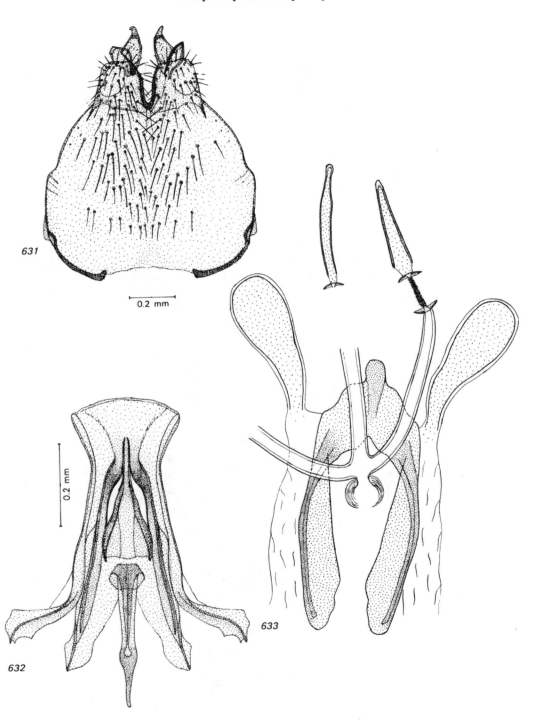

Figs. 631–633: *Exoprosopa suffusa*
631. gonopods; 632. aedeagus; 633. spermatheca

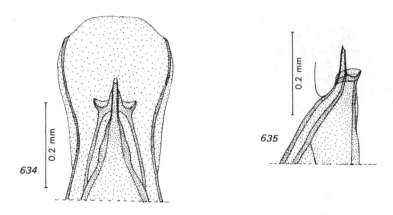

Figs. 634–635:   *Exoprosopa bagdadensis*
634. apex of aedeagus; 635. same, lateral

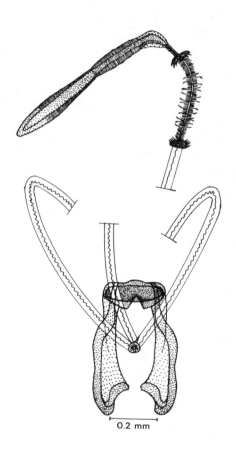

Fig. 636:   *Exoprosopa tamerlan bezzii*, spermatheca

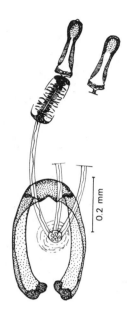

Fig. 637:   *Exoprosopa* (*Pterobates*) *chalybea*, spermatheca

species. However, species with similar spermathecae differed distinctly in coloration and wing venation. Species with short, tubular spermathecae (type 2) were placed by Bowden in the genus *Heteralonia* Rondani which was established for a species with an aberrant wing venation (*oculata* Macquart). The differences between *Heteralonia* and *Exoprosopa* given by Bowden (1975b) are not convincing and he states (p. 371) that the former genus is 'difficult to define'. The male genitalia do not differ markedly from those of other species of *Exoprosopa* except in specific characters. It seems advisable to restrict *Heteralonia* to the type species *oculata* and the closely related species *kaokoensis* Hesse, 1956, as done by Hull (1973). *E. pygmalion*, which has an aberrant wing venation, has spermathecae of type 2. Furca rectangularly U- or V-shaped, with specific differentiations in most species. Tergite 8 either without an apodeme but with only a strongly sclerotized anterior margin, or with a very wide, short apodeme which extends with two ridges into the tergite, or the apodeme is divided into two triangular parts (*chalybea*). Tergite 9 with 3–5 thick spines in most species, 6–7 in a few; spines very thick in *minos*.

E. telamon is usually placed in *Exoprosopa* because of the presence of three submarginal cells, but Engel (1936) stated that it is probably a *Thyridanthrax*. This is supported by the structure of the gonocoxites with proximally directed processes which resemble those of the subgenus *Exhyalanthrax* of *Thyridanthrax*.

Two Australian species examined have tubular spermathecae but with specific differences. The spermathecae of *E. fenestrata* resemble those of *Micromitra* described below also in the special structure of the ejection apparatus. *E. fenestrata* was placed by Roberts in the genus

223

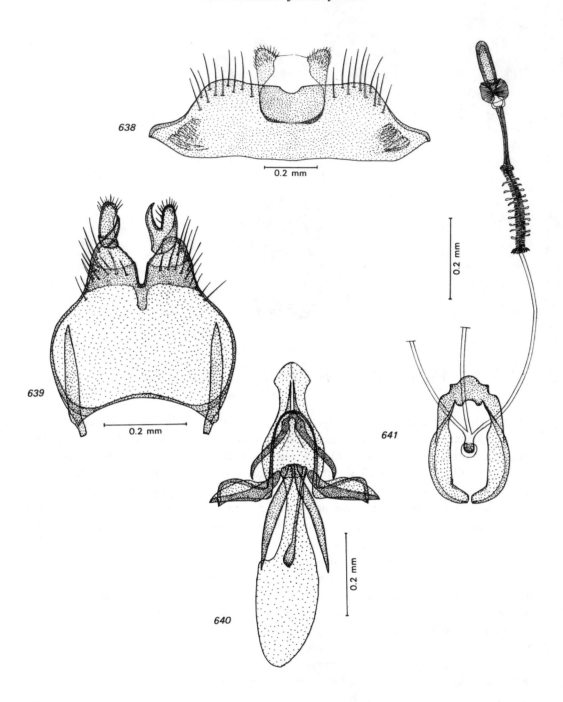

Figs. 638–641: *Exoprosopa* sp. no. 1
638. epandrium; 639. gonopods; 640. aedeagus; 641. spermatheca

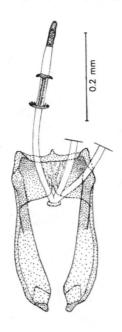

Fig. 642: *Exoprosopa pygmalion*, spermatheca

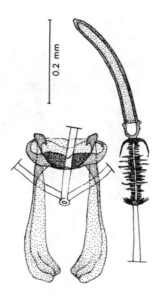

Fig. 643: *Exoprosopa latelimbata*, spermatheca

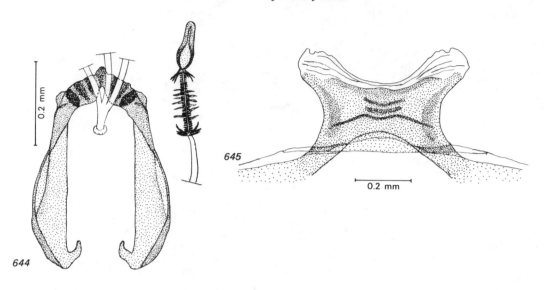

Figs. 644–645: *Exoprosopa divisa*
644. spermatheca; 645. tergite 8 of female, apodeme

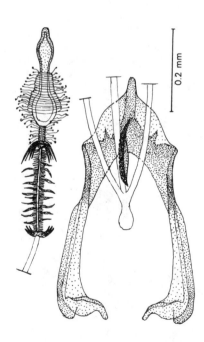

Fig. 646: *Exoprosopa rostrifera*, spermatheca

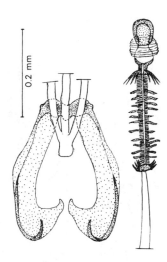

Fig. 647: *Exoprosopa caliptera*, spermatheca

Pseudopenthes, but it may have to be placed in the genus *Micromitra* which it also resembles in some external characters.

The male genitalia of the four American species examined resemble those of the Palaearctic species in general. Epandrium of varying form, with relatively small, rounded median part and long, pointed basal lateral processes in *caliptera*. Dististyli more or less bifid. Aedeagal process long, with rounded apex, more or less wide, with specific differences. It has a large ventral bulge in *rostrifera*.

The spermathecae differ distinctly from those of the Palaearctic species. The capsules of *divisa* are tubular apically and widened in the basal part. Ducts to vagina extremely long. The capsules of *rostrifera* are widened in the middle, with a short, narrow, tubular apex. They continue in a short, sclerotized duct to the ejection apparatus. This duct is absent in the other three species. End plates of ejection apparatus with long processes. *E. doris* has small, conical capsules and *caliptera* has rounded capsules. Apodeme of tergite 8 of varying form, broad and projecting in *divisa* and *doris*, very short and wide in *caliptera* and *rostrifera*.

The genitalia, particularly those of the female, thus show a surprising variation, but most differences are apparently of specific rank and cannot be used in order to form species groups. However, the differences in external characters and in the genitalia are so marked in some cases that the creation of new genera seems justified. The *stupida* group has therefore been made the genus *Micromitra* by Bowden. Generic rank may also be justified for some other groups, e.g., the *tamerlan* group and the subgenus *Pterobates*.

227

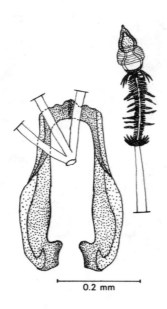

0.2 mm

Fig. 648: *Exoprosopa doris*, spermatheca

Micromitra Bowden, 1964

M. iris, pharao, (?) *stupida*, and an undescribed species (Figs. 649–652)

The species differ from those of *Exoprosopa* in the iridescent coloration of the scales, the narrow vertex and the male and female genitalia.

Bowden (1964) stated in the diagnosis of the genus that the hairs on the thorax and abdomen of the African species are pectinate. This is not the case in the Palaearctic species examined, in which all hairs are simple.

Epandrium rounded posteriorly, with short, pointed basal lateral corners. Gonocoxites fused. Dististyli with long, narrow, curved apical part and broad, triangular base so that they appear half-ring-shaped. Aedeagus very short, with wide, rounded basal part. Apodeme long and narrow. Aedeagal process with wide apex with lateral corners (? *stupida*), or rectangular, with two rounded apical processes with denticles (*pharao*).

Capsules of spermathecae forming long, uniformly wide, recurved tubes. Ejection apparatus very short, without processes, with a large, broadly cup-shaped basal end plate and a similar, smaller apical plate. Its duct is divided in the middle by a short, membranous part. Furca U-shaped, with an apical process of specific form. Tergite 8 without apodeme, but with strongly sclerotized middle of the anterior margin. Tergite 9 with two groups of 4–5 strong, slightly curved spines.

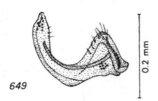

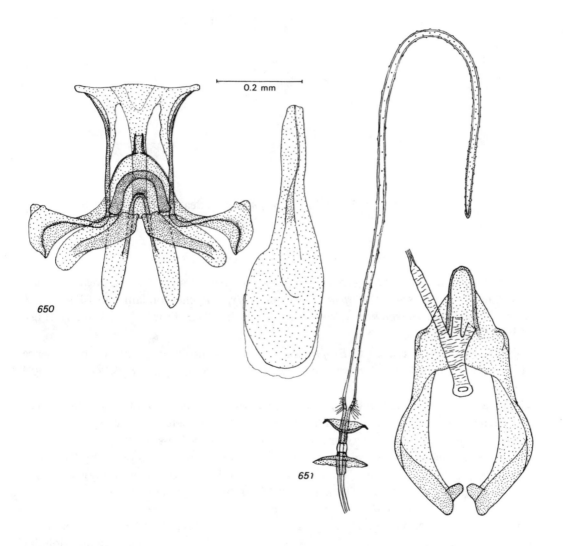

Figs. 649–651: *Micromitra* (?) *stupida*
649. dististylus; 650. aedeagus; 651. spermatheca

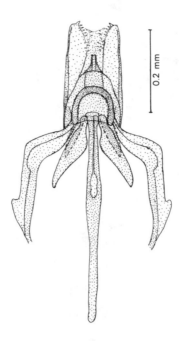

Fig. 652: *Micromitra pharao*, aedeagus

Ligyra Newman, 1841 (**Hyperalonia** auct.)

L. ferrea (Palaearctic); *L. gazophylax, servillei* (Nearctic); *L. mars, venus, vittata* (Ethiopian); *L. chrysolampris, sphinx* (Oriental); *Hyperalonia* Rondani sensu Painter, *H. chilensis, morio, surinamensis* (Neotropical); *L. cingulata, satyrus* (Australian) (Figs. 653–661)

This genus is closely related to *Exoprosopa* from which it differs mainly in the presence of four submarginal cells. It was considered as a subgenus of *Exoprosopa* by Engel (1936).

The male genitalia of the Ethiopian species examined resemble those of *Exoprosopa* except for specific differences in the form of the dististyli and of the aedeagal process. Illustrations of the genitalia of *vittata, mars* and others were given by Hesse (1956).

Several Neotropical species (*chilensis, morio, surinamensis*) were considered by Painter as the separate genus *Hyperalonia* Rondani, but this is probably a subgenus of *Ligyra* and was so considered by Hull (1973). The male genitalia do not differ significantly from those of the Old World species of *Ligyra*, but the spermathecae are different.

The spermathecae of the Old World species examined are of type 1 of *Exoprosopa* but there are exceptions, e.g., *L. vittata*. The spermathecae of this species have small, globular capsules and very long ducts which are at first narrow, then become wider and form a widening at the base. This has an internal sclerotization and continues in a very narrow duct to the ejection apparatus which has a double apical end plate and a larger basal end

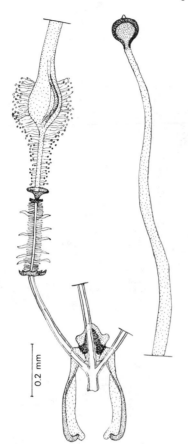

Fig. 653: *Ligyra vittata*, spermatheca

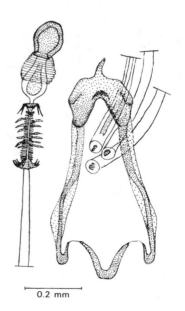

Fig. 654: *Ligyra mars*, spermatheca

231

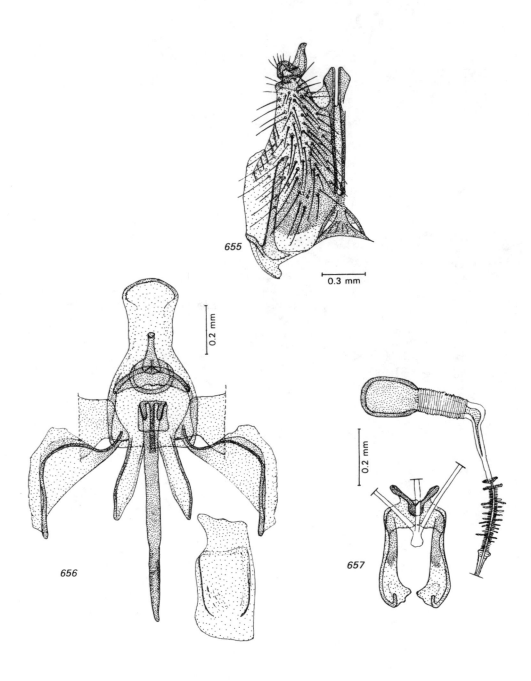

Figs. 655–657: *Ligyra (Hyperalonia) morio*
655. gonopod; 656. aedeagus; 657. spermatheca

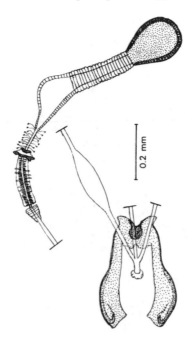

Fig. 658: *Ligyra (Hyperalonia) chilensis*, spermatheca

plate with pointed processes. Furca rectangularly U-shaped, with an apical process having three rounded corners. A second specimen identified as *vittata* has larger, oblong-oval capsules, the internal sclerotization in the base of the ducts is absent, and the apical process of the furca is different. This is possibly a different species.

Tergite 8 with a wide, very short apodeme in all species, except in *L. satyrus* which has a long, wide, T-shaped apodeme.

The species of the subgenus *Hyperalonia* have different spermathecae. The capsules are large, oblong-oval or globular, continuing in a wide striated duct which widens proximally and then narrows into a narrow duct to the ejection aparatus. This has small and longer processes and a double apical end plate. Ducts to vagina widened before the ending. Basal end plate absent. Furca rectangularly U-shaped, with two diverging apical processes. Tergite 8 with a narrow apodeme which differs distinctly from the wide, very short apodeme of the other species. Most Old World species examined have a V-shaped furca which also differs distinctly from that of *Hyperalonia*

Comparison of the genitalia of *Ligyra gazophylax* and *servillei* with those of species of *Hyperalonia* showed their great similarity, particularly of the spermathecae. The differences are only of specific rank, so that the position of *Hyperalonia* as a subgenus of *Ligyra* seems justified.

The genitalia of *L. ferrea* show some special differentiations. The gonocoxites have proximally directed processes in the apical part, like species of *Exhyalanthrax*, and the dististyli are bifid. The aedeagal process has a pointed, curved apical process.

233

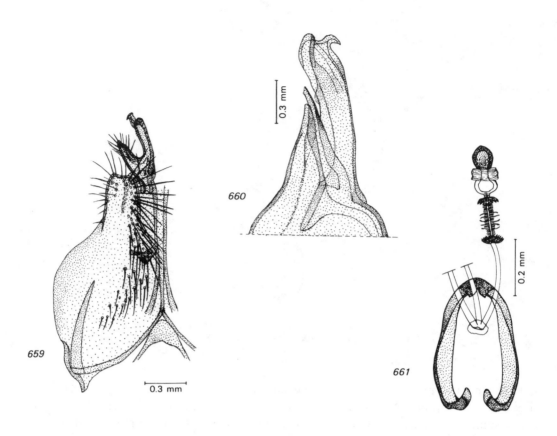

Figs. 659–661: *Ligyra ferrea*
659. gonopod; 660. aedeagus, apical part; 661. spermatheca

Litorhina Bowden, 1975 (= **Litorrhynchus** Macquart, 1840)

L. basalis, erythraea, nyasae, obumbrata, pseudocellata (Figs. 662–667)

This genus is also closely related to *Exoprosopa* from which it differs mainly in its long proboscis. The male genitalia resemble those of *Exoprosopa*. The aedeagal process shows specific differentiations in the species examined.

The spermathecae are different from those of *Exoprosopa*. The capsules are tubular, short, more or less wide, with a rounded apex which is slightly wider than the basal part in some species. There is a wide, folded membranous part at the base of the capsule. Ejection apparatus short, with short and longer processes and large end plates with pointed processes. Furca U- or V-shaped, with specific differentiations. Tergite 8 with a short, wide apodeme. Tergite 9 with 4–6 strong spines.

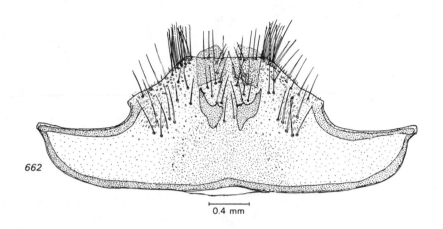

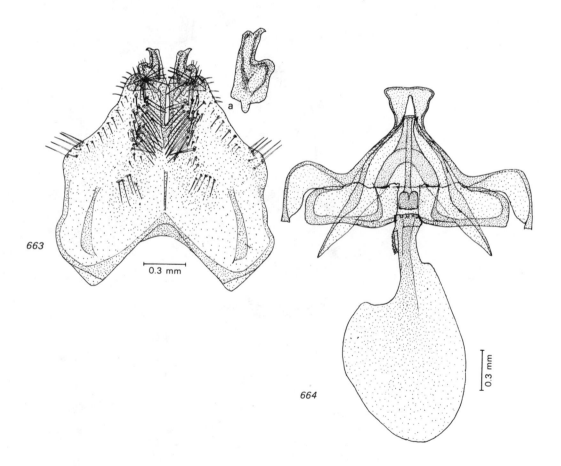

Figs. 662–664 : *Litorhina erythraea*
662. epandrium; 663. gonopods; (a) dististylus, lateral; 664. aedeagus

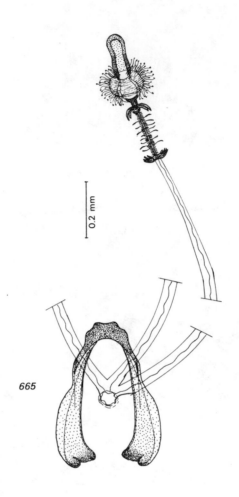

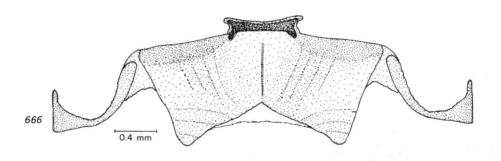

Figs. 665–666:   *Litorhina erythraea*
665. spermatheca; 666. tergite 8 of female

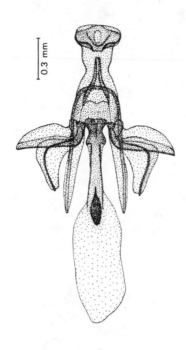

Fig. 667: *Litorhina obumbrata*, aedeagus

Thyridanthrax Osten-Sacken, 1886

(Figs. 668–705)

This large genus contains forms with very different wing coloration. The subgenus *Exhyal-anthrax* Becker (type species *vagans* Loew) was based mainly on the clear wings. This subgenus could be retained, but requires redefinition according to the characteristic structure of the male and female genitalia as it also contains species with a distinct dark wing pattern.

The type species of the genus is the Nearctic species *selene* Osten-Sacken which has not been examined. Most species have two submarginal cells, but a few (*ternarius, telamon*) have three such cells. Some species placed in *Thyridanthrax*, e.g., *T. latona*, differ so distinctly from the others that they may have to be placed in another genus.

There are two distinct groups in the Palaearctic species of the genus according to the structure of the genitalia: the *polyphemus* group and the subgenus *Exhyalanthrax*. A nominate subgenus cannot be established at present as the American type species *selene* may belong to a different species group.

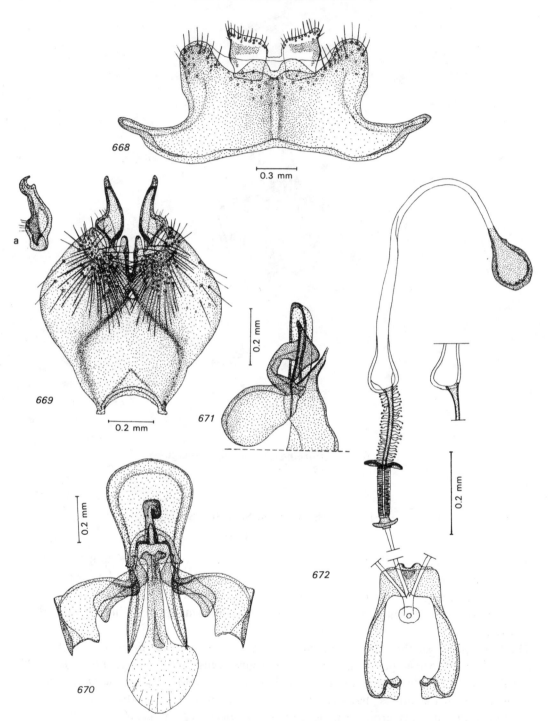

Figs. 668–672: *Thyridanthrax perspicillaris*
668. epandrium; 669. gonopods; (a) dististylus, lateral; 670. aedeagus;
671. same, apical part, lateral; 672. spermatheca

Polyphemus Group

T. elegans, griseolus, incanus, misellus, obliteratus, perspicillaris, polyphemus, punctum, ternarius, telamon, and several undescribed species (Palaearctic); *T. pallidus, pertusus* (Nearctic) (Figs. 668–680)

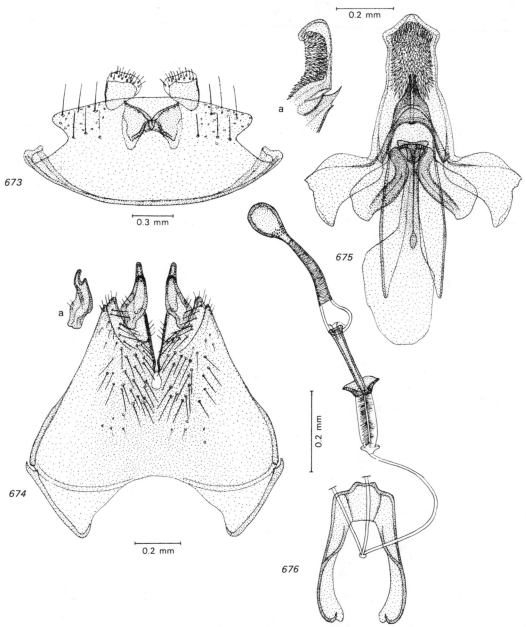

Figs. 673–676: *Thyridanthrax elegans*
673. epandrium; 674. gonopods; (a) dististylus, lateral; 675. aedeagus;
(a) apex, lateral; 676. spermatheca

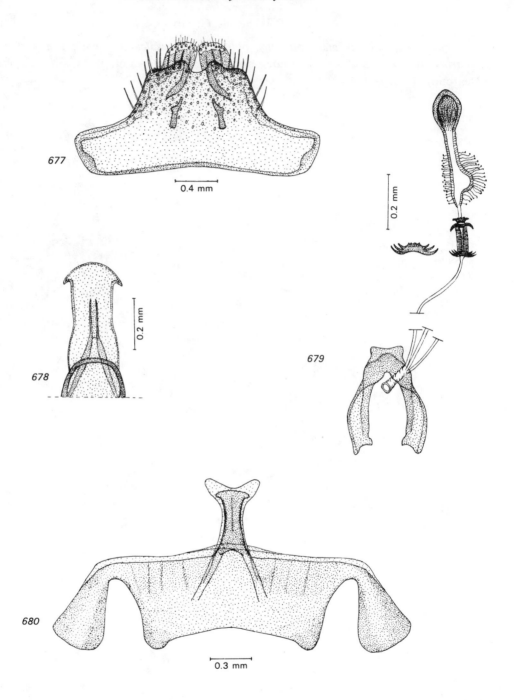

Figs. 677–680: *Thyridanthrax punctum*
677. epandrium; 678. aedeagus, apical part; 679. spermatheca; 680. tergite 8 of female

The epandrium is of the same type as in *Exoprosopa* but differs in form and in the length of its basal lateral processes in the various species. These processes are usually shorter than those in *Exoprosopa* and their sides are sometimes distinctly concave. The subanal plates of the proctiger (sternite 10 or 11) are often of distinctive form. Gonocoxites truncate-triangular, often more or less completely fused in the basal part. Dististyli triangular, curved, or with a tubercle before the pointed apex. Aedeagus conical; aedeagal process flattened, broad, rounded apically or densely covered with denticles and of distinctive form (*incanus, elegans*). It has a special differentiation in *perspicillaris* and *ternarius*—a thick, loop-shaped structure in the middle. Aedeagal process of *T. punctum* with two pointed lateral processes at the apex.

The spermathecae have rounded or ovoid capsules and a long or very long, narrow duct which is asymmetrically widened at the base. Then follows a narrow, sclerotized duct to the ejection apparatus which is strongly sclerotized and has large end plates. Furca U- or V-shaped. Tergite 8 without an apodeme in most species, only with a short, strongly sclerotized part in the middle of the anterior margin. *T. punctum*, which belongs to this group according to most characters, has a distinct, narrow, T-shaped apodeme on tergite 8, resembling that of species of *Exhyalanthrax*. Tergite 9 with 3–4 spines on each side.

T. perspicillaris and *ternarius* are very similar in most characters, particularly in the structure of the aedeagus, but *ternarius* has three submarginal cells. The two species differ in their distribution. *T. perspicillaris* is Mediterranean, while *ternarius* is mainly Ethiopian but it occurs in Egypt, Sinai and southern Israel (Negev) where the distribution of the two forms overlaps. The presence of two or three submarginal cells is thus apparently not a generic character in all cases. Engel (1936) considered *ternarius* as a subspecies of *perspicillaris*.

Subgenus **Exhyalanthrax** Becker, 1916

Revised diagnosis. Wings with or without a dark pattern.
Gonocoxites of male with proximally directed processes bearing setae in the apical part. Aedeagal process broadly rounded, usually with two denticles in the apical part.
Spermathecae of type 1 of *Exoprosopa*, with small rounded capsules, short ducts which widen proximally, and a membranous part below the capsule. Apodeme of tergite 8 distinctly projecting, broad, extending with ridges into the body of the tergite.

T. afer, agnitionalis, lepidulus, perpusillus, vagans, and several undescribed species (Figs. 681–691)

Epandrium more or less rectangular, with rounded posterior corners and short basal lateral processes. Gonocoxites partly or completely fused, with a proximally directed, more or less long process bearing numerous setae in the apical part of each gonocoxite. Dististyli triangular, pointed or with a tubercle before the apex. Aedeagus conical, more or less long and wide. Aedeagal process wide, flattened, usually with a broadly rounded apex and with or without two lateral denticles near the apex.
The spermathecae resembly type 1 of *Exoprosopa*, with small, rounded or oblong capsules, a broad, membranous, striated part below the capsule, and a short sclerotized duct to the

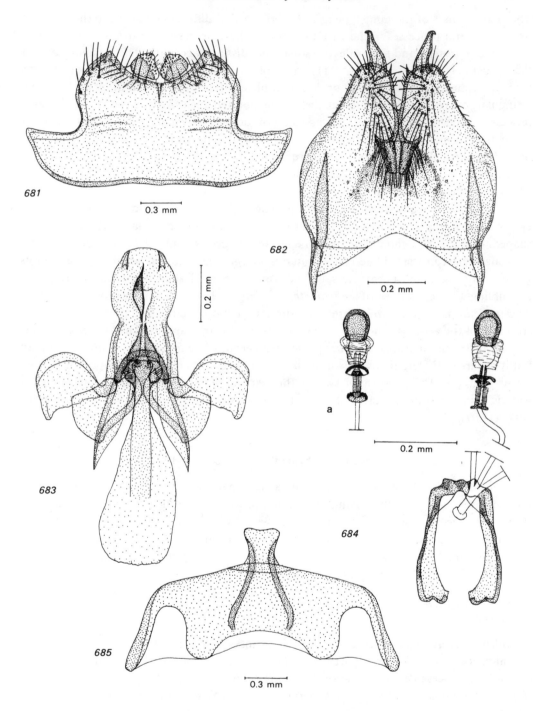

Figs. 681–685: *Thyridanthrax (Exhyalanthrax) vagans*
681. epandrium; 682. gonopods; 683. aedeagus; 684. spermatheca;
(a) spermatheca of *T. (Exhyalanthrax) agnitionalis*; 685. tergite 8 of female of *T. vagans*

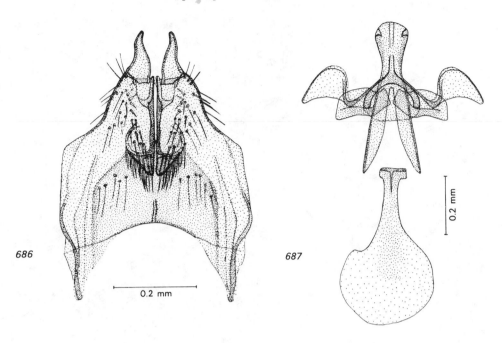

Figs. 686–687: *Thyridanthrax (Exhyalanthrax) perpusillus*
686. gonopods; 687. aedeagus

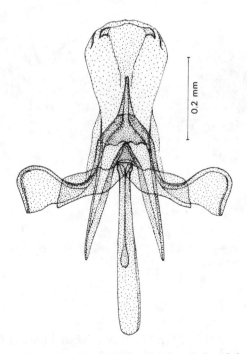

Fig. 688: *Thyridanthrax (Exhyalanthrax) lepidulus*, aedeagus

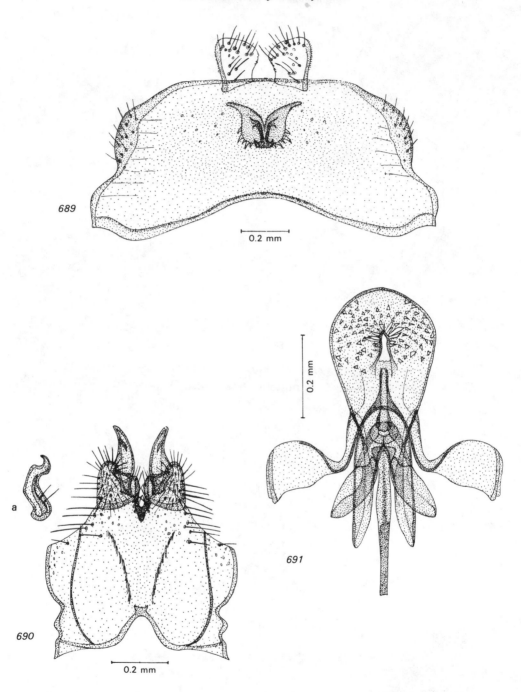

Figs. 689–691: *Thyridanthrax (Exhyalanthrax) fulvifacies*
689. epandrium; 690. gonopods; (a) dististylus, lateral; 691. aedeagus

244

ejection apparatus which is short, with large end plates and only small processes. Furca rectangularly U-shaped. Tergite 8 with a moderately broad, long apodeme which extends with two ridges into the body of the tergite. Tergite 9 with 4–6 short spines on each side, only two spines in *perpusillus*.

T. latona, fulvifacies, telamon, pallidus, pertusus (Figs. 692–705)

Several species do not fit into the above two groups. *T. latona,* which may have to be removed from the genus, has no processes on the gonocoxites which are divided. Aedeagal process broadly rounded, with a triangular process at the apex. The spermathecae are distinctly different. Capsules oblong, rhomboidal, pointed.

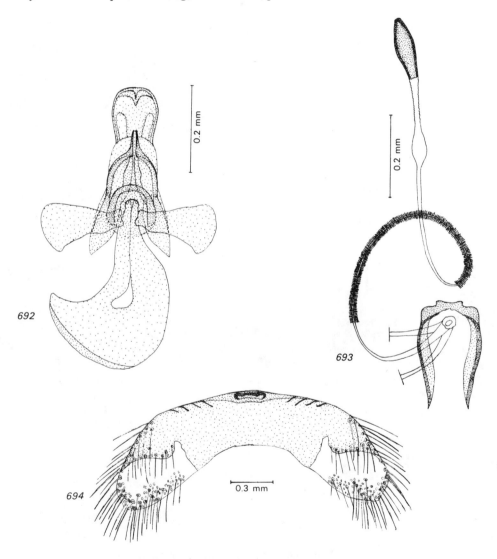

Figs. 692–694: *Thyridanthrax latona*
692. aedeagus; 693. spermatheca; 694. tergite 8 of female

245

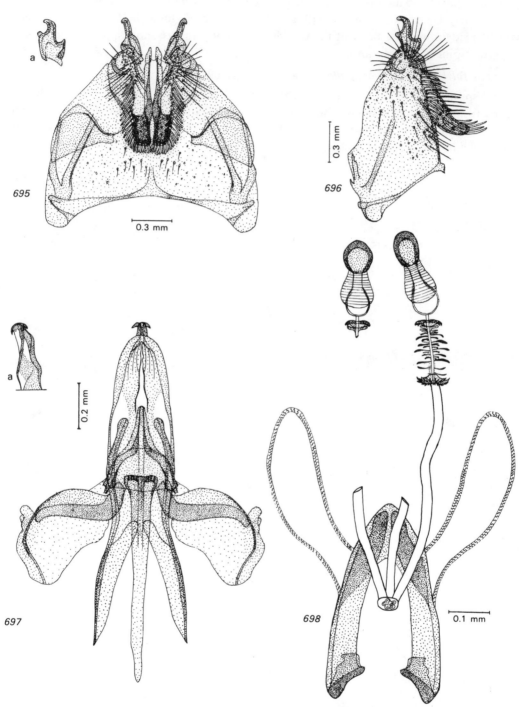

Figs. 695–698: *Thyridanthrax telamon*
695. gonopods; (a) dististylus, lateral; 696. gonopod, lateral; 697. aedeagus;
(a) apex of aedeagal process, lateral; 698. spermatheca

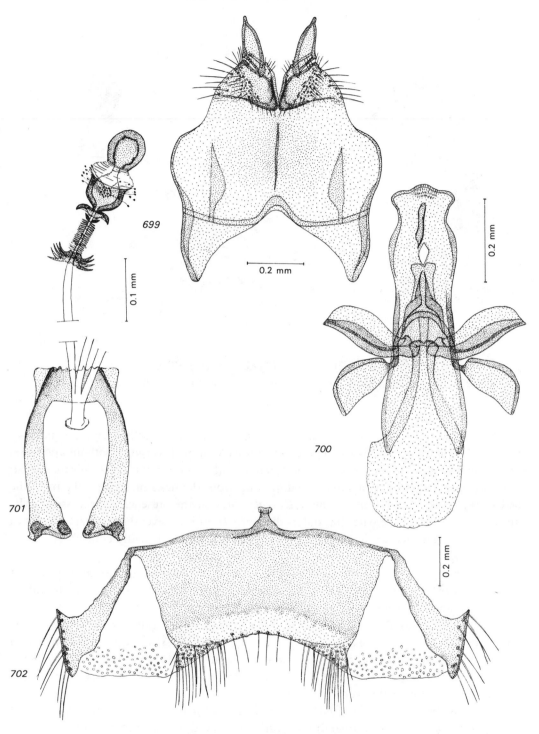

Figs. 699–702 : *Thyridanthrax* sp. no. 1
699. gonopods; 700. aedeagus; 701. spermatheca; 702. tergite 8 of female

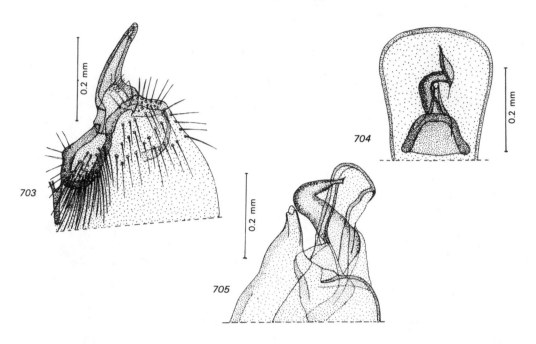

Figs. 703–705: *Thyridanthrax pallidus*
703. apex of gonopod; 704. apex of aedeagus, dorsal; 705. same, lateral

Ducts very long, at first wide, then narrowing. Ejection apparatus very long, narrow, lacking end plates, with only small processes. Furca V-shaped. Tergite 8 without apodeme, only with a more strongly sclerotized median part of the anterior margin. *T. fulvifacies* also has no processes on the gonocoxites and has numerous denticles on the apical part of the aedeagal process. *T. telamon* resembles *Exhyalanthrax* in the presence of processes on the apical part of the gonocoxites but differs distinctly from it in several characters. Dististyli with two short processes in the basal part. Aedeagus short, broadly conical. Aedeagal process long, tapering, with a small, curved, bifid apical process.

Spermathecae transitional between those of the *polyphemus* group and those of *Exhyalanthrax*, with small, rounded capsules and a short, striated duct which widens proximally and is surrounded by a membranous structure.

Species no. 1 differs distinctly from the other species of the genus in the form of the antennae — with a long style as in some species of *Exoprosopa*, in the wing venation, and in the spermathecae with a basal sclerotization in the membranous part below the capsule. Ducts to ejection apparatus absent. Tergite 8 of the female with a short, narrow apodeme. It may not belong to this genus.

The two American species examined, *T. pallidus* and *pertusus*, also apparently do not belong to the above groups. Epandrium with long, curved, pointed, basal lateral processes and deeply concave sides in *pallidus*, sides less concave in *pertusus*. Gonocoxites with short apical processes and a thick brush of long setae at the apex in *pallidus*, without processes

248

and setae in *pertusus*. Aedeagal process in both species broadly rounded, with a loop-shaped structure resembling that described for *perspicillaris*.

Spermathecae with rounded, slightly flattened capsules, a membranous part behind the capsule which is intermediate between a duct and the wide structure in type 1 of *Exoprosopa*, and a relatively long sclerotized duct to the ejection apparatus which has a large apical and a smaller basal end plate and only short processes. Furca rectangularly U-shaped. Tergite 8 without apodeme, only with a strongly sclerotized median part of the anterior margin. Tergite 9 with five spines on each side in *pallidus*, with only two spines in *pertusus*. Tergite 8 of *pertusus* with a short, wide apodeme.

Chrysanthrax crocina Osten-Sacken, 1892
(Figs. 706, 707)

Epandrium short, trapezoidal, with short, broad, rounded, projecting posterior corners. Basal lateral processes short, pointed, curved. Gonocoxites fused; dististyli with rounded base and long, pointed apical part. Aedeagus short, its base bulging. Aedeagal process with broadly rounded apex, without differentiations. Hall (1975) gives lateral views of the genitalia of Chilean species of the genus, some of which show differentiations of the apex of the aedeagal process.

Spermathecae with small, rounded capsules and moderately wide, short ducts with an asymmetrical proximal widening as in *Thyridanthrax*. There is a short, sclerotized duct to the ejection apparatus which is short, with small processes, a double apical end plate and a basal plate transformed into a transparent funnel as in *Paravilla*. Ducts to vagina long, narrow. Furca rectangularly U-shaped. Tergite 8 with a narrow apodeme. Tergite 9 with 2–3 long spines on a small process on each side.

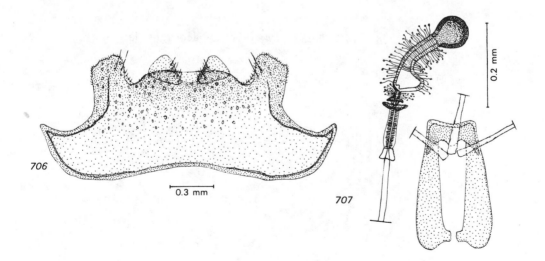

Figs. 706–707: *Chrysanthrax crocina*
706. epandrium; 707. spermatheca

Villa Lioy, 1864

V. abbadon, atricauda, bivirgata, clarissima (= *noscibilis* Austen), *decipula, hottentotta, insignis, melanura, senecio*, and several undescribed species (Palaearctic); *V. aenea, lateralis, pretiosa* (Nearctic) (Figs. 708–719)

This genus is differently defined by the various authors. All Palaearctic species examined have a rounded, not projecting face, but American authors placed *Thyridanthrax, Chrysanthrax* and other genera as subgenera in the genus *Villa*. Only species with a rounded face are here considered to belong to this genus.

Epandrium more or less trapezoidal, with short basal lateral processes but with specific differences in form. Gonocoxites truncate-triangular, fused in the basal part. Dististyli triangular or more or less parallel-sided and with a tubercle before the pointed apex. Hypandrium absent, apparently fused with the gonocoxites. Aedeagus conical. Aedeagal process broad, with apex either rounded or divided into a rounded median process and two narrow, pointed lateral processes, often with or without a ventral, more or less long hook.

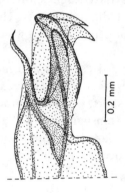

Fig. 708:   *Villa hottentotta*,   apex of aedeagus, lateral

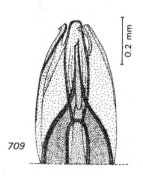
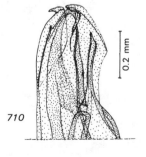

Figs. 709–710:   *Villa insignis*
709. apex of aedeagus, dorsal; 710. same, lateral

250

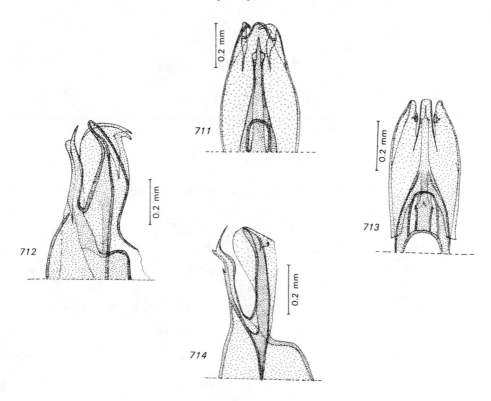

Figs. 711–714: *Villa bivirgata*
711. type 1, apex of aedeagus, dorsal; 712. same, lateral;
713. type 2, apex of aedeagus, dorsal; 714. same, lateral

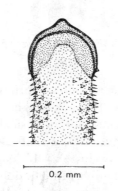

Fig. 715: *Villa decipula*, apex of aedeagal process

251

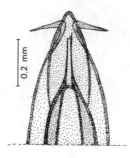

Fig. 716:   *Villa pretiosa*, apex of aedeagus

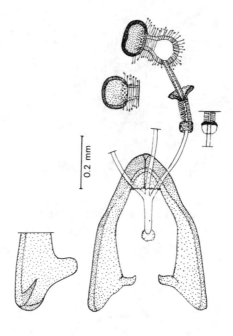

Fig. 717:   *Villa clarissima*, spermatheca

Spermathecae with rounded capsules and a short, wide, membranous duct which is asymmetrically widened proximally. This continues in a relatively long, narrow, sclerotized duct to the ejection apparatus which is short, with small end plates and small processes or only with an apical end plate. Tergite 8 with a relatively long, narrow apodeme. Furca U- or V-shaped. Tergite 9 with two groups of 4–12 long spines.

The American species examined show only specific differences from the above description. The capsules of *V. pretiosa* are of different form, the ducts are much longer, and its aedeagal process has two long, lateral apical processes.

V. decipula shows a number of differences. The aedeagal process is broadly rounded and covered with small denticles. Tergite 8 lacks an apodeme, but has only a more strongly sclerotized median part of the anterior margin. This species also differs from the others examined in having a distinct dark wing pattern resembling that of *Thyridanthrax incanus*.

The aedeagal process of *V. abbadon* shows some distinct differences. It has two rounded apical processes and a semicircular group of spines before the apex (Fig. 718).

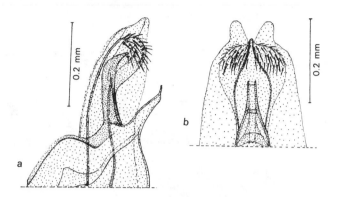

Fig. 718 : *Villa abbadon* ; (a) apex of aedeagus, lateral ; (b) same, dorsal

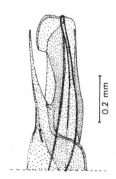

Fig. 719: *Villa senecio*, apex of aedeagus

Villa atrata Coquillett, V. miscella Coquillett
(Figs. 720–724)

In the past these two species were placed in the genus *Villa*. Stone (1965) considered *V. atrata* as belonging to the subgenus *Thyridanthrax* of *Villa*. They differ distinctly from species of *Villa* as considered here in the form of the face and the antennae, in the wing coloration and in the male and female genitalia. Dr J. Hall informed me that he intends to establish a new genus for them and this seems justified.

Epandrium of *V. atrata* trapezoidal, with rounded posterior corners and short, pointed

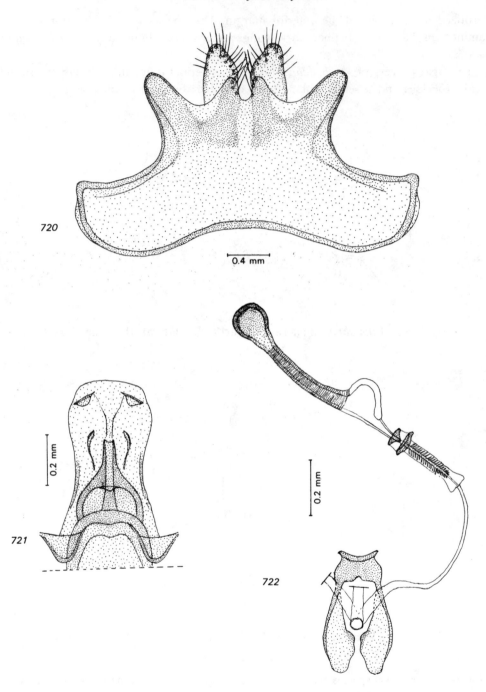

Figs. 720–722: *Villa miscella*
720. epandrium; 721. aedeagus; 722. spermatheca

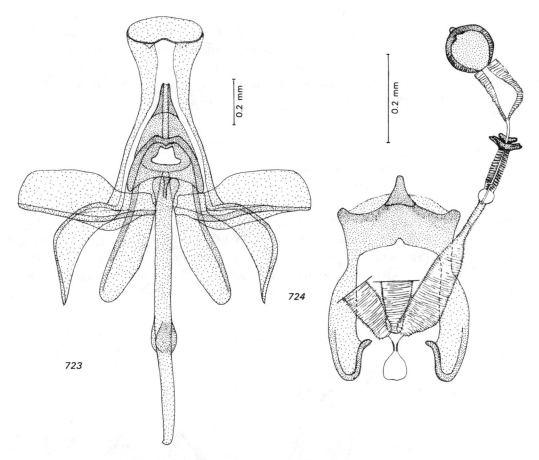

Figs. 723–724: *Villa atrata*
723. aedeagus; 724. spermatheca

lateral basal processes, with distinctly obliquely projecting posterior lateral corners and two small processes in the middle of the posterior margin and with deeply concave sides in *V. miscella*. Gonocoxites triangular, with a large brush of long setae on the apical part in *miscella*. Dististyli with broad, projecting basal part and narrow, pointed apical part. Hypandrium small, fused with the gonocoxites. Aedeagus broadly conical, short, with bulging base. Aedeagal process broad, rounded, with two small, rounded processes in the apical part in *miscella*, rectangular, with a transverse ridge in *atrata*.

Spermathecae of *atrata* with globular, thick-walled capsules and a small apical process. Ducts very short, wide, striated, continuing in a narrow duct to the ejection apparatus which is very short, with a double apical end plate and short processes. Basal end plate transformed into a transparent rounded structure. Ducts to vagina at first narrow, then broadly widening and striated to near the vagina. Furca rectangularly U-shaped, with a triangular apical process. Tergite 8 with a broad, long, slightly T-shaped apodeme. Tergite 9 with four thick spines on each side.

Spermathecae of *V. miscella* with smaller capsules with a conical base. Ducts long, with an asymmetrical proximal widening as in species of *Thyridanthrax*. Ejection apparatus also with a double apical end plate, the basal end plate forming a transparent funnel (Fig. 722). Ducts to vagina narrow, not striated, only slightly widening proximally. Furca rectangularly U-shaped, with two lateral apical processes. Tergite 8 with a broad apodeme but not T-shaped. Tergite 9 with four thick spines on each side.

Hemipenthes Loew, 1869

H. hamifera, morio, velutinus (Figs. 725–732)

This genus resembles *Villa* in the rounded, not projecting face and differs from it mainly in the extensively black wings, the reduced patagium and other characters.

Epandrium trapezoidal; lateral basal processes more or less short. Gonocoxites fused, truncate-triangular. Dististyli triangular or with a tubercle before the apex. Aedeagus short, conical, with a large basal bulge. Aedeagal process broad, slightly flattened, with very small denticles in the greater apical part in *velutinus*, narrower, narrowing proximally,

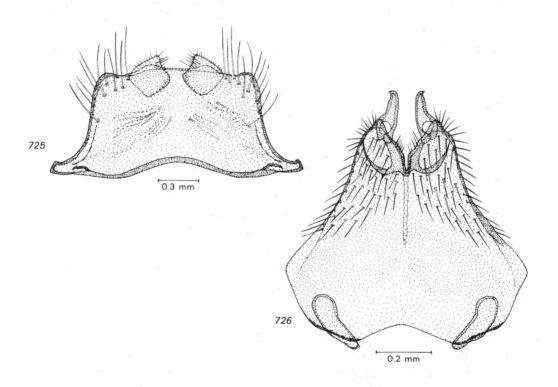

0.3 mm

0.2 mm

Figs. 725–726: *Hemipenthes hamifera*
725. epandrium ; 726. gonopods

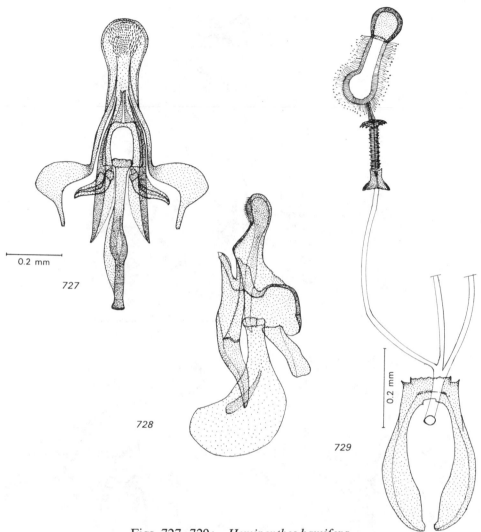

0.2 mm

727

728

0.2 mm

729

Figs. 727–729: *Hemipenthes hamifera*
727. aedeagus, dorsal; 728. same, lateral; 729. spermatheca

with more distinct denticles in the apical part in *hamifera*, with a large apical median denticle and small denticles in the apical part and along the sides in *morio*.

Spermathecae with small rounded capsules or capsules with a conical base and with short, wide ducts having an asymmetrical basal widening. Ducts to ejection apparatus short, sclerotized. Ejection apparatus short; apical end plate moderately large; basal plate transformed into a funnel-shaped, nearly transparent structure as in *Lepidanthrax, Paravilla* and other American genera. Furca rectangularly U-shaped, with specific differences. Tergite 8 with a short, relatively broad, truncate apodeme, or longer and narrower and slightly T-shaped. Tergite 9 with 4–7 short, thick spines with curved apex on each side.

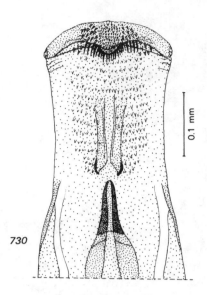

Fig. 730: *Hemipenthes morio*, aedeagal process, apex

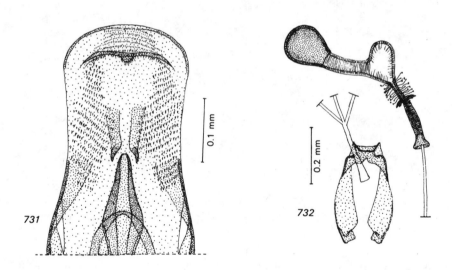

Figs. 731–732: *Hemipenthes velutinus*
731. aedeagal process, apex; 732. spermatheca

Paravilla cinerea Cole, 1823
(Figs. 733–736)

This genus was also considered as a subgenus of *Villa* by Stone (1965) and Painter, Painter & Hall (1978), but it differs so distinctly from *Villa* both in external characters and in the

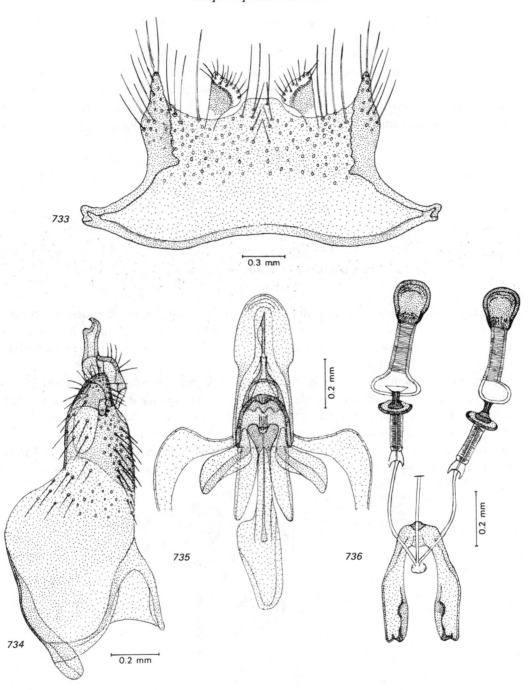

Figs. 733–736: *Paravilla cinerea*
733. epandrium ; 734. gonopod ; 735. aedeagus ; 736. spermathecae

genitalia that it seems justified to maintain the genus. It resembles *Villa* in the form of the patagium at the base of the wings.

Epandrium of distinctive form, broadly rectangular, with long, pointed posterior lateral processes and a broad process in the middle of the posterior margin. Lateral basal processes moderately long, pointed. Gonocoxites fused; dististyli with broadly rounded basal part and curved, pointed, narrow apical part with a subapical tubercle. Aedeagus short, with broadly rounded base. Aedeagal process with rounded apex, slightly narrowing basally and with a longitudinal ridge, but without differentiations.

Spermathecae with small rounded capsules having a conical base. Ducts short, wide, with an asymmetrical basal widening. Duct to ejection apparatus very short, sclerotized. Ejection apparatus short, with a double apical end plate and a basal end plate transformed into a funnel-shaped transparent structure. Furca U-shaped. Tergite 8 with a broad, truncate apodeme. Tergite 9 with five very thick, short spines on each side.

Oestanthrax brunnescens (Loew, 1857)

(Fig. 737)

Epandrium trapezoidal, without distinct lateral basal processes. Gonocoxites fused, truncate-triangular. Dististyli triangular, curved, with a subapical tubercle. Aedeagus conical; basal plates very small. Aedeagal process broad, with rounded apex and a median ridge, lacking differentiations.

Spermathecae resembling those of *Villa*, with rounded capsules and a short, wide, membranous part below the capsule. Duct to ejection apparatus sclerotized, striated. Ejection apparatus short, its apical end plate with processes, the basal plate transformed into a transparent, rounded structure. Furca rectangularly U-shaped, with three apical tubercles. Tergite 8 with a short, moderately wide apodeme. Tergite 9 with two groups of about 10 strong spines.

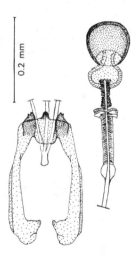

Fig. 737: *Oestanthrax brunnescens*, spermatheca

Poecilanthrax Osten-Sacken, 1885

P. autumnalis, sackeni (Figs. 738, 739)

Epandrium trapezoidal, with short lateral basal processes. Gonocoxites fused or incompletely divided, narrowly triangular. Dististyli triangular, with a subapical tubercle. Aedeagus conical; aedeagal process simple, with rounded apex, of varying form according to the illustrations given by Painter & Hall (1960).

Spermathecae of characteristic form. Capsules rounded, more or less flattened. Ducts very short, widening into a long, sausage-shaped extension which reaches, without an intermediate duct, to the ejection apparatus. This is short, with small processes and large, cup-shaped end plates with an apical process. Ducts to vagina striated, widening proximally. Furca U-shaped, rectangular. Tergite 8 with a large, T-shaped apodeme. Tergite 9 with 4–6 large, curved spines on each side.

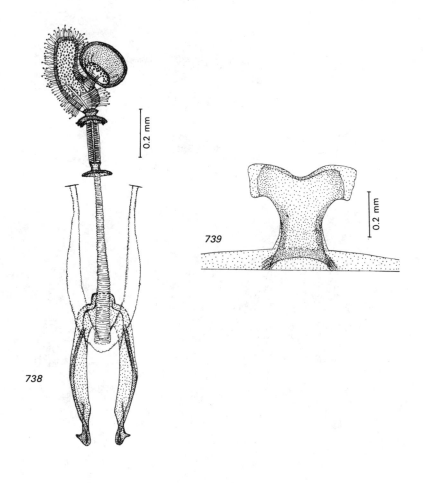

Figs. 738–739: *Poecilanthrax sackeni*
738. spermatheca; 739. apodeme of tergite 8 of female

Cyananthrax cyanoptera (Wiedemann, 1830)

(Figs. 740–744)

Epandrium short, trapezoidal, with rounded posterior corners and long basal lateral processes. Gonocoxites divided, triangular. Dististyli triangular, with an angle before the apex. Aedeagus conical. Aedeagal process long, with rounded apex, lacking differentiations.

Spermathecae resembling those of *Poecilanthrax*. Capsules flattened, with an asymmetrical, long, membranous part below the capsule and with a short, sclerotized duct to the ejection apparatus which has two small end plates without processes and with small processes on the duct. Furca rectangularly U-shaped, with a T-shaped process at the apex. Tergite 8 with a large, T-shaped apodeme. Tergite 9 with two groups of six large spines.

The genitalia of both sexes show a distinct resemblance to those of *Poecilanthrax*.

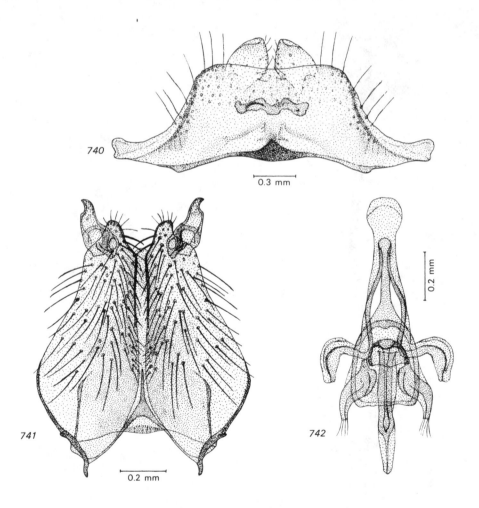

Figs. 740–742: *Cyananthrax cyanoptera*
740. epandrium; 741. gonopods; 742. aedeagus

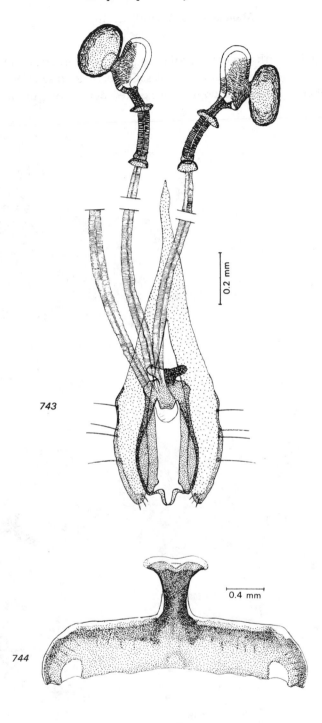

Figs. 743–744: *Cyananthrax cyanoptera*
743. spermathecae; 744. tergite 8 of female

Mancia nana Coquillett, 1886

(Figs. 745–748)

Epandrium rectangular, with projecting, rounded posterior corners, an indentation in the middle of the posterior margin and concave sides. Gonocoxites fused. Dististyli triangular, curved, with a tubercle before the pointed apex. Aedeagus short, broadly conical, with

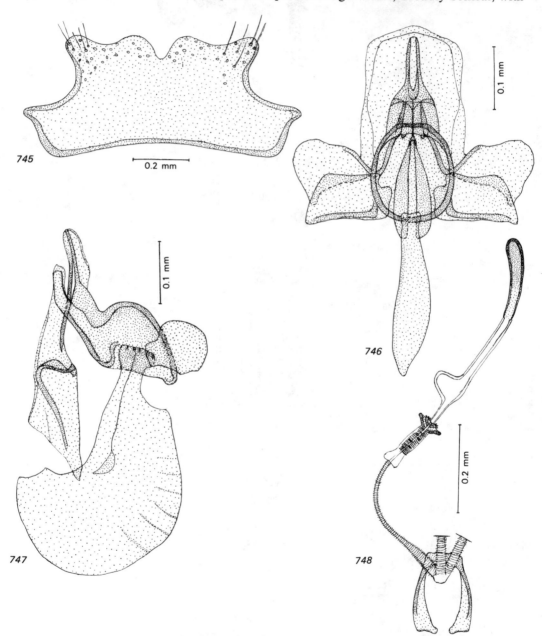

Figs. 745–748: *Mancia nana*

745. epandrium; 746. aedeagus; 747. same, lateral; 748. spermatheca

bulging base. Aedeagal process short, very wide, with rounded apex, without differentiations. Apodeme large, triangular, curved.

Spermathecae with short, tubular capsules with rounded apex. Ducts moderately long and wide, with an asymmetrical basal widening. Ejection apparatus short, with short processes. Apical end plate double in one specimen, its apical part with denticles; however, this part is absent in a second specimen examined. Basal plate transformed into a transparent funnel. Furca rectangularly U-shaped. Tergite 8 with a narrow apodeme with a triangular base. Tergite 9 with two very short spines in the specimen with a single end plate, without spines in the specimen with a double apical end plate.

Dipalta serpentina Osten-Sacken, 1877

(Figs. 749–751)

Epandrium rectangular, with short basal lateral processes. Gonocoxites fused, triangular. Dististyli with broad basal part and long, parallel-sided apical part with a sharp angle before the apex. Aedeagus short, with very wide base. Aedeagal process broad, with rounded apex, lacking differentiations.

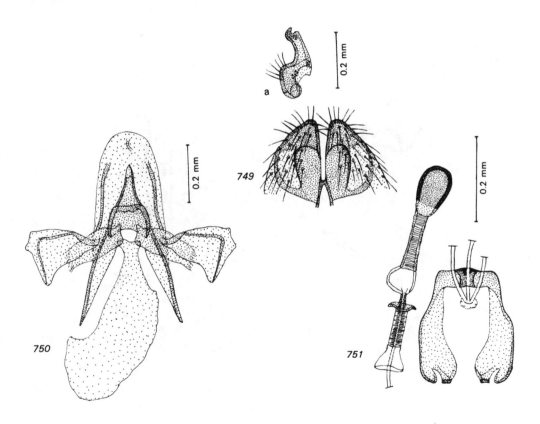

Figs. 749–751: *Dipalta serpentina*
749. apex of gonopods; (a) dististylus; 750. aedeagus; 751. spermatheca

Spermathecae with oblong-ovoid capsules; ducts short, wide, with a proximal widening. Duct to ejection apparatus sclerotized. Ejection apparatus with small processes, a single apical end plate and a basal plate transformed into a large, transparent funnel. Furca rectangularly U-shaped. Tergite 8 with a narrow apodeme. Tergite 9 with two groups of three strong spines.

Astrophanes adonis Osten-Sacken, 1886
(Figs. 752, 753)

Epandrium short, trapezoidal, with short, broad, basal lateral processes. Gonocoxites fused, triangular. Dististyli triangular, with a sharp angle in the basal part. Aedeagus conical. Aedeagal process wide, with rounded apex, lacking differentiations.

Spermathecae with narrow, oblong-oval capsules; ducts narrow, short, with a proximal widening. Duct to ejection apparatus narrow, sclerotized. Ejection apparatus short, with short processes, a small, simple apical end plate and a basal plate transformed into an irregularly rounded membranous structure. Furca U-shaped, with two small processes at the apex. Tergite 8 with a short, narrow apodeme. Tergite 9 with two groups of six strong spines.

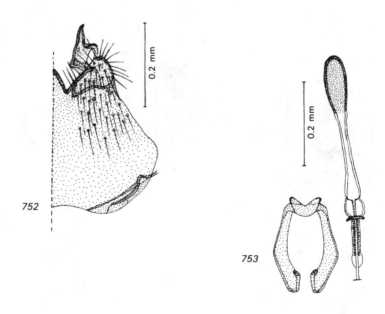

Figs. 752–753: *Astrophanes adonis*
752. gonopod; 753. spermatheca

Diplocampta secunda Paramonov, 1931
(Figs. 754–756)

Epandrium trapezoidal, with short, pointed, basal lateral corners. Gonocoxites fused, triangular. Dististyli triangular, with a tubercle bearing hairs on the broad basal part. Aedeagus with long, narrow apical part and relatively narrow basal part. Aedeagal process long, relatively narrow, with slightly rounded apex, lacking differentiations.

Spermathecae with short, nearly cylindrical capsules having a rounded apex. Ducts long, with a basal widening; duct to ejection apparatus short, sclerotized. Ejection apparatus with short processes, a single apical end plate and a basal end plate transformed into a conical membranous structure. Furca rectangularly U-shaped. Tergite 8 with a long, narrow apodeme. Tergite 9 with two groups of three thin setae.

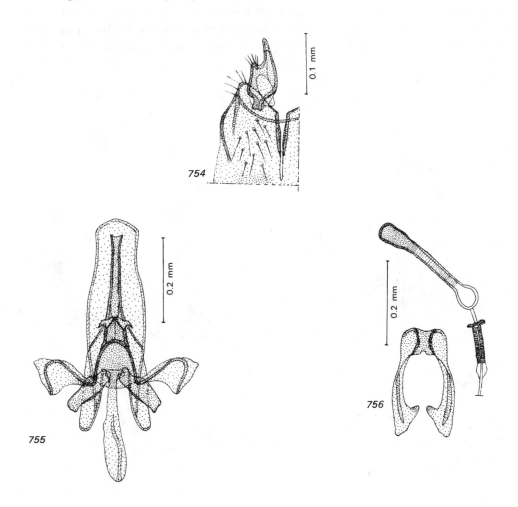

Figs. 754–756: *Diplocampta secunda*
754. apex of gonopods; 755. aedeagus; 756. spermatheca

Neodiplocampta Curran, 1934

N. mira, sepia (Figs. 757–761)

Epandrium trapezoidal in *sepia*, curved, with longer, pointed lateral basal processes, with short, rounded, projecting posterior corners and more or less concave sides in *mira*. Gonocoxites fused, with broad basal and narrow apical part. Dististyli of *mira* with rounded basal part and curved, narrow apical part with a subapical tubercle. Dististyli of *sepia* nearly parallel-sided, curved, also with a subapical tubercle. Aedeagus short, with wide, rounded base. Aedeagal process very broad, rounded, with a rounded apical process which bears a long, pointed, curved process in *mira*, with only a median ledge and a few denticles at the apex in *sepia*.

Spermathecae with small, club-shaped capsules in *mira*, with broadly pear-shaped capsules in *sepia*. Ducts wide, with an asymmetrical basal widening. Ejection apparatus short, with short processes, a wide apical end plate and a basal plate transformed into a transparent funnel. Furca rectangularly U-shaped. Tergite 8 with a short, narrow apodeme in *sepia*, with a broadly T-shaped apodeme bearing apical processes in *mira*. Tergite 9 with two groups of 3–4 curved spines in *mira*, 5–6 in *sepia*.

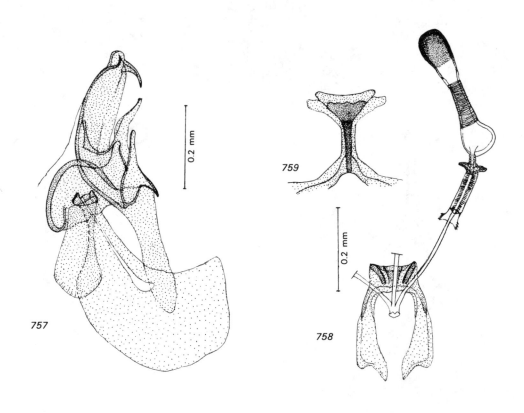

Figs. 757–759: *Neodiplocampta mira*
757. aedeagus; 758. spermatheca; 759. apodeme of tergite 8 of female

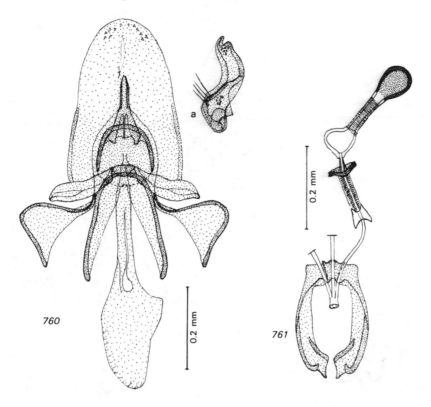

Figs. 760–761: *Neodiplocampta sepia*
760. aedeagus; (a) dististylus; 761. spermatheca

Lepidanthrax Osten-Sacken, 1886

L. agrestis, proboscideus (Figs. 762–768)

Epandrium trapezoidal, with broad, truncate posterior lateral processes, concave sides and moderately long, basal lateral processes in *agrestis*, curved, with longer basal lateral processes, a deeply indented posterior margin and concave sides in *proboscideus*. Gonocoxites fused, triangular in *agrestis*. Dististyli narrow, nearly parallel-sided, with a subapical tubercle in *agrestis*. Gonocoxites of *proboscideus* similar, but with a dense group of long, black setae in the apical part. Dististyli curved, with broader basal part than in *agrestis*. Aedeagus conical, with broadly rounded base. Aedeagal process wide, nearly rectangular, with a broad, short median apical process in *agrestis*. Aedeagal process of *proboscideus* similar, but longer and narrower, with two very small, laterally directed apical processes.

Spermathecae of *agrestis* with short, tubular, slightly club-shaped capsules with rounded apex. Ducts as wide as the capsules, with an asymmetrical basal widening and a short striated duct to the ejection apparatus. This is short, with a wide apical end plate and a basal plate transformed into a transparent funnel. Furca V-shaped. Tergite 8 with a short, narrow apodeme.

269

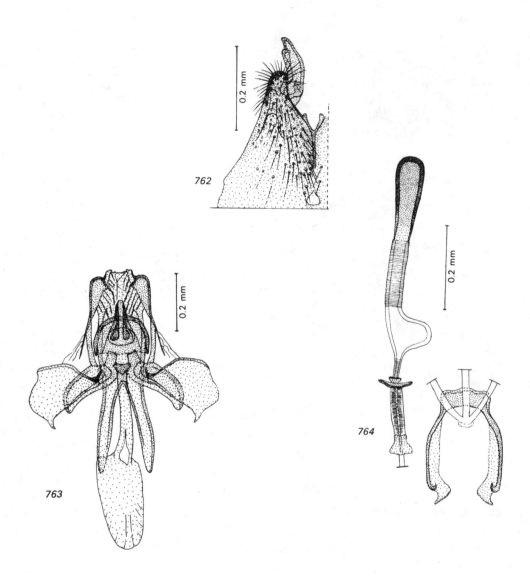

Figs. 762–764: *Lepidanthrax agrestis*
762. apex of gonopods; 763. aedeagus; 764. spermatheca

Spermathecae of *proboscideus* distinctly different. Capsules rounded, broadly oval, with a conical apical inner process. Ducts short, wide, their basal widening much larger. Ducts to ejection apparatus very short, sclerotized. Ejection apparatus as in *agrestis*, but apical end plate with processes and the funnel-shaped basal plate with sclerotized basal part. Furca U-shaped, with a tubercle at the basal ends. Tergite 9 divided, with 4–6 short, straight spines on each part in both species.

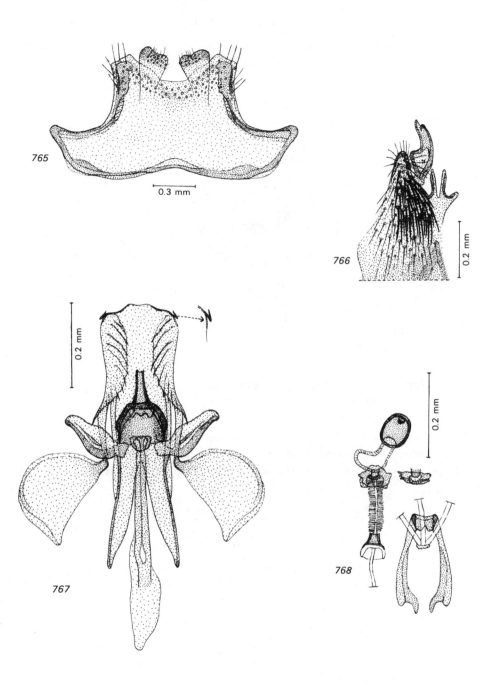

Figs. 765–768: *Lepidanthrax proboscideus*
765. epandrium; 766. apex of gonopods; 767. aedeagus; 768. spermatheca

Prorates Melander, 1906

P. anomalus, (?) *bezzii*, and several undescribed species (Palaearctic); *P. frommeri* and an undetermined species (Nearctic) (Figs. 5–7, 20–23, 769)

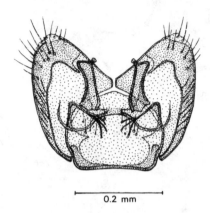

0.2 mm

Fig. 769: *Prorates* (?) *bezzii*, genitalia, dorsal

As stated above (pp. 8, 18), this genus probably belongs to the Scenopinidae.

Tergite 2 of the abdomen has a sensory area with short spines and hairs.

Epandrium of male divided, parts truncate-triangular (not divided in all Bombyliidae examined). Gonocoxites fused into a nearly rectangular sclerite, with posterior processes of varying form. Dististyli of varying form, triangular or nearly parallel-sided and curved, with a small apical process in *frommeri*. Aedeagus described on p. 6–8, spermathecae on p. 18.

CONCLUSIONS

THE MALE GENITALIA of the Bombyliidae cannot be used for the identification of pinned specimens, except in a few cases. They are small, usually retracted and often covered with dense hairs. Dissection of the male genitalia is therefore necessary. All parts—epandrium, gonopods, and particularly the form of the aedeagal process in dorsal view—provide definite specific characters. Most authors gave lateral views of the complete genitalia, but this does not permit recognition of many important characters of the aedeagus complex.

The structure of the female genitalia, particularly of the spermathecae, has not been used in the past. Only recently Marston (1970), in his revision of the American species of *Anthrax*, gave illustrations of the female genitalia, and Mühlenberg (1970, 1971) described the morphology of the posterior segments of the female and the spermathecae of a number of species.

Attempts to divide some large genera (*Bombylius, Exoprosopa, Thyridanthrax, Usia*) into subgenera or species groups, mainly on the basis of wing venation and coloration, have not been successful, and many new species have been described according to minor variations of these characters which have proved very variable in some species when longer series or specimens from different localities were examined. In some cases distinctly different genitalia were found in specimens which apparently belonged to the same species according to external characters (*Geron, Petrorossia, fuscipennis* group of *Anthrax*).

The insufficiency of external characters alone was also demonstrated in the case of the subgenus *Exhyalanthrax* of *Thyridanthrax*. This subgenus was based by Becker (1916) on the clear wings. Examination of the characteristic genitalia showed that its retention is justified, but the same genitalia were also found in species with a distinct, black wing pattern. A revised definition of the subgenus is therefore given.

The spermathecae, the sclerites behind the furca and tergite 8 provide important systematic characters. The spermathecae of the Bombyliidae are varied and differentiated to even a greater degree than those of the Asilidae as described by Theodor (1976). The spermathecae of some genera (*Glabellula, Doliopteryx*) show unusual differentiations, the function of which is not clear (Figs. 43, 46). The spermathecae show generic characters in some genera (*Geron, Heterotropus, Lomatia, Usia, Phthiria*), but their variation is apparently mainly specific. Thus six different types of spermathecae were found in Palaearctic species of *Exoprosopa*, only two of which appear in a larger number of species, the others only in one or two species; additional types were found in American species. Two or three different types of spermathecae were found in *Amictus* and *Anastoechus*. The *ater* group of *Bombylius* is well defined by external characters, but four different types of spermathecae were found in the six species of the group examined. They can therefore not be used for the formation of species groups or subgenera.

The American material examined made it possible in some cases to determine whether genera recorded both from the Old and the New World belonged to the same genus. In certain cases the American species differed distinctly from Palaearctic species of the genus and may have to be considered as subgenera or even as different genera. Some genera have been wrongly recorded from America, e.g., *Dischistus* which occurs only in the Ethiopian and Mediterranean regions. The same applies apparently to *Cyllenia*. Painter, Painter & Hall (1978) considered *Sphenoidoptera* as a synonym of *Cyllenia*. However, the genitalia of *Sphenoidoptera* illustrated by Hull (1973) differ so distinctly from those of *Cyllenia* that the synonymy seems unjustified.

Hull (1973, p.438) suggested that there is a trend of simplification of the wing venation in the Bombyliidae. A reduced wing venation is found mainly in the very small species (*Empidideicus, Cyrtosia* and related genera) and the most complicated wing venation mainly in the large species (*Ligyra*). This is apparently connected with the size of the wings, as large wings need additional veins to ensure stability. An example of this was given by Rensch (1959, p.210) for species of Chrysomelidae: the wing venation of a small species of a group is distinctly reduced in comparison with a large species.

A phylogenetic scheme cannot yet be constructed on the basis of our present knowledge. It would have to be based on the study of much more extensive material, including the greater part of the genera of the family and description of the external characters and the genitalia of both sexes.

If all the information discussed above is taken into account, it seems justified to maintain the family Bombyliidae in its present composition and not to divide it into different families, as all groups show some common characters and there are transitions between them. The division into Homoeophthalmae and Tomophthalmae should be abandoned, and the family should be divided into a reduced number of subfamilies and tribes which should be defined by a combination of external characters and the structure of the genitalia of both sexes.

BIBLIOGRAPHY

Austen E. E. (1937) 'Bombyliidae of Palestine', *Brit. Mus. (Nat. Hist.)*.

Becker T. (1906) '*Usia* Latreille', *Berl. Ent. Zeitschr.*, Vol. 1.

— (1916) 'Beiträge zur Kenntnis einiger Gattungen der Bombyliiden', *Ann. Mus. Hung.*, Vol. 14.

Berlese A. (1909) Gli Insetti, Vol. 1, Soc. Ed. Libr. Milano.

Bezzi M. (1924) 'The Bombyliidae of the Ethiopian Region', *Brit. Mus. (Nat. Hist.)*.

Bowden J. (1964) 'The Bombyliidae of Ghana', *Mem. Ent. Soc. South Africa*, No. 8.

— (1971) 'Notes on Some Australian Bombyliidae in the Zoological Museum in Copenhagen. *Steenstrupia*', *Zool. Mus. Copenhagen*, Vol. 1.

— (1974) 'Studies on African Bombyliidae. VIII. On the Geroninae', *J. Ent. Soc. South Africa*, Vol. 37, No. 1.

— (1975a) 'IX. On *Hyperusia* Bezzi and the Tribe Corsomyzini', *J. Ent. Soc. South Africa*, Vol. 38, No. 1.

— (1975b) 'X. Taxonomic Problems Relevant to a Catalogue of Ethiopian Bombyliidae', *J. Ent. Soc. South Africa*, Vol. 38, No. 2.

Crampton G. C. (1942) 'Guide to Insects of Connecticut', Part VI, Diptera, *State Geol. Hist. Surv. Bull.* No. 64.

Efflatoun H. C. (1945) 'A Monograph of Egyptian Diptera. VI. Bombyliidae', *Bull. Soc. Fouad I d'Entomologie*, Vol. 29.

Engel E. O. (1932–1937) in: Lindner, *Die Fliegen der Palaearktischen Region.* 25. Bombyliidae. Stuttgart, Schweizerbart.

Hall J. C. (1969) 'A Review of the Subfamily Cylleniinae and a World Revision of the Genus *Thevenemyia* Bigot', *Univ. Calif. Publ. Ent.*, Vol. 56.

— (1972) 'New North American Heterotropinae', *Pacif. Entom.*, Vol. 48.

— (1975) 'The Bombyliidae of Chile', *Pacif. Entom.*, Vol. 76.

Hesse A. J. (1938, 1956) 'A Revision of the Bombyliidae of Southern Africa', *Ann. South African Mus.*, Vols. 34, 35.

Hull F. M. (1973) *Bee Flies of the World*, Smithson. Inst. Press, Wash., D.C.

Kelsey L. P. (1969) 'A Revision of the Scenopinidae of the World', *U.S.N.M. Bull.* 277.

Marston N. (1963) 'A Revision of the Nearctic Species of the *albofasciatus* Group of the Genus *Anthrax* Scopoli', Agr. Exp. Stat., Kansas State Univ., *Techn. Bull.* 127.

— (1970) 'Revision of the New World Species of *Anthrax* other than the *A. albofasciatus* Group', *Smithson. Contr. Zool.*, No. 43.

Melander A. L. (1927) *Genera Insectorum: Diptera Fam. Empididae.* Fasc. 185.

— (1950) 'Taxonomic Notes on Some Smaller Bombyliidae', *Pan-Pacific Entom.*, Vol. 26.

Mühlenberg M. (1970) 'Besonderheiten im Bau der Receptacula seminis von parasitischen Fliegen (Bombyliidae)', *Zool. Jb. Anat.*, Vol. 87.

— (1971) 'Die Abwandlung des Eilegeapparates der Bombyliidae', *Z. Morph. Tiere*, Vol. 70.

Painter R. H. & J. C. Hall (1960) 'A Monograph of the Genus *Poecilanthrax*', Agr. Exp. Stat. Kansas State University, *Techn. Bull.* 106.

Painter R. H. & E. M. Painter (1962) 'Redescriptions of Types of North American Bombyliidae in European Museums', *J. Kansas Entom. Soc.*, Vol. 35.

— (1963) 'A Review of the Subfamily Systropinae in North America', *J. Kansas Entom. Soc.*, Vol. 36, No. 4.

Painter R. H., E. M. Painter & J. C. Hall (1978) 'Catal. Diptera of America South of the U.S. 38. Bombyliidae', Museum São Paulo.

Paramonov S. J. (1928) 'Beiträge zur Monographie der Gattung *Exoprosopa*', *Trav. Mus. Zool. Kiev*, No. 4.

— (1929) 'Beiträge zur Monographie einiger Bombyliiden Gattungen', *Trav. Mus. Zool. Kiev*, No. 6.

Pendergrast J. G. (1957) 'Studies on Reproductive Organs of the Heteroptera with Consideration of their Bearing on Classification', *Trans. Roy. Ent. Soc. Lond.*, Vol. 109, Part 1.

Rensch B. (1959) *Evolution above the Species Level*, Methuen, London.

Roberts F. H. S. (1928) 'A revision of the Australian Bombyliidae', I–II, *Proc. Linn. Soc. Sydney*, Vol. 54.

Rohdendorf B. B. (1964) Inst. Akad. Nauk SSSR, Vol. 100, Moscow.

Séguy E. (1938) Étude sur les Dipteres recueillis par M. H. Lhote dans le Tassili des Ajjer', *Encycl. Ent.*, Paris, Ser. B. Dipteres, Vol. 9. p.34.

Stone A. (1965) 'Catalogue of the Diptera of America North of Mexico', Agric. Handbook No. 276.

Theodor O. (1976) *On the Structure of the Spermatheca and Aedeagus in the Asilidae and their Importance in the Systematics of the Family*, Israel Academy of Sciences and Humanities, Jerusalem.

Weber H. (1933) *Lehrbuch der Entomologie*, Fischer, Jena.

Zaitzev V. F. (1969) *Keys to the Insects of the W. European U.S.S.R.*, Vol. V, Part 1, Akad. Nauk SSSR, Leningrad.

BIBLIOGRAPHY

סודר ונדפס בדפוס כתר, ירושלים

כתבי האקדמיה הלאומית הישראלית למדעים

החטיבה למדעי־הטבע

איברי המין של זבובי הפרחים

מאת

א. תאודור

ירושלים תשמ״ג